ROUTLEDGE LIBRARY EDITIONS: WATER RESOURCES

Volume 6

LIFE BEFORE THE DROUGHT

LIFE BEFORE THE DROUGHT

Edited by
EARL SCOTT

LONDON AND NEW YORK

First published in 1984 by Allen & Unwin Inc.

This edition first published in 2024
by Routledge
4 Park Square, Milton Park, Abingdon, Oxon OX14 4RN

and by Routledge
605 Third Avenue, New York, NY 10158

Routledge is an imprint of the Taylor & Francis Group, an informa business

British Library Cataloguing in Publication Data
A catalogue record for this book is available from the British Library

ISBN: 978-1-032-74502-2 (Set)
ISBN: 978-1-032-74481-0 (Volume 6) (hbk)
ISBN: 978-1-032-74491-9 (Volume 6) (pbk)
ISBN: 978-1-003-46945-2 (Volume 6) (ebk)

DOI: 10.4324/9781003469452

Publisher's Note
The publisher has gone to great lengths to ensure the quality of this reprint but points out that some imperfections in the original copies may be apparent.

Disclaimer
The publisher has made every effort to trace copyright holders and would welcome correspondence from those they have been unable to trace.

LIFE BEFORE THE DROUGHT

Edited by
Earl Scott

Boston
ALLEN & UNWIN
London Sydney

Allen & Unwin Inc.,
9 Winchester Terrace, Winchester, Mass. 01890, USA

George Allen & Unwin (Publishers) Ltd,
40 Museum Street, London WC1A 1LU, UK

George Allen & Unwin (Publishers) Ltd,
Park Lane, Hemel Hempstead, Herts HP2 4TE, UK

George Allen & Unwin Australia Pty Ltd,
8 Napier Street, North Sydney, NSW 2060, Australia

First published in 1984

Library of Congress Cataloging in Publication Data

Main entry under title:
Life before the drought.
Includes bibliographical references.
1. Man—Influence of environment—Sahel—Addresses, essays, lectures. 2. Human ecology—Sahel—Addresses, essays, lectures. 3. Sahel—Economic conditions—Addresses, essays, lectures. I. Scott, Earl. II. Title: Savanna–Sahel zones of Africa.
GF740.L53 1984 333.7 84-9235
ISBN 0-04-910076-9 (cased)

British Library Cataloguing in Publication Data

Life before the drought.
1. Africa, West–Social conditions
I. Scott, Earl P.
966 HN820.A8
ISBN 0-04-910076-9

Set in 10 on 12 point Bembo by Preface Ltd, Salisbury
and printed in Great Britain by Mackays of Chatham

For Rebecca, Stephanie and Lynne

Preface

This book is about people and how they adjust to a harsh (my perception) environment. A focus on people, as opposed to a 'climatic region' or an 'uncommon event', requires an interest in social history, inter-ethnic relations, resource endowment and management, and factors influencing standards of living. This book presents these interests by considering life in the Savanna–Sahel zones prior to the uncommonly disastrous drought of 1968–74 and by drawing attention to the relationship between poverty and famine, due to the loss of purchasing power or to the loss of entitlements to food, as opposed to a catastrophic event or the unavailability of food. The purpose of the book is to provide information on how Africans dealt with their environment and how they managed their resources so that their strategies, at least the proven ones, could be considered for incorporation into any plan for economic and ecological recovery of the Savanna–Sahel zones.

The book developed out of the editor's interest in folk knowledge and land-use patterns in northern Nigeria. The intent was to compile an interdisciplinary reader, employing a human ecological approach, on the pastoral and agricultural economies employed in the Savanna–Sahel zones of Africa before the drought of 1968–74. The belief was that farmers and herders throughout this arid and semi-arid region had devised strategies for coping with an uncertain and harsh environment, and that these strategies, though born in the past, were applicable to the future. The 'Sahel' and the 'Drought' were not viewed as appropriate reference points to begin a serious discussion of life and poverty in the Savanna–Sahel zones of Africa. Both prescribe unwarranted limits to a search for causes and solutions. Instead, we must begin with the people and their access to available resources of land, water, and food.

I am pleased to have this opportunity to thank the people who helped to make this book a reality. I am especially grateful to the authors for their patience when unforeseen events diverted my attention from this book. I wish to thank Mary Chisley and Djenaba El-Shabazz of the Department of Afro-American and African Studies and Rebecca Hogan and Margaret Rasmussen of the Department of Geography for their long and untiring efforts to type the manuscript; and the cartographic staff of the Department of Geography, for their production of the illustrations in this book. I would also like to express my regrets to Professor J. W. Ssennyonga, whose work on 'Managerial and demographic aspects of the indigenous irrigation systems of the Marakmet of Kenya' could not appear in this volume, but for whose willingness to contribute I am quite grateful.

EARL P. SCOTT

Contents

List of tables

1 *Introduction: life and poverty in the Savanna–Sahel zones*

EARL P. SCOTT

Much has been written concerning the drought that ravaged the Savanna–Sahel zones[1] of West Africa from 1968 to about 1974. Most of the concern centered on the drought itself, its cause(s), its intensity, its extent, and its possible duration. The most jarring revelation came with reports that over 100 000 people had perished from hunger, needlessly (Morris & Sheets 1974, p. 134). In response to this, international conferences were convened not only to exchange information on the Sahelian drought but also to discuss a related and potentially more devastating phenomenon, desertification. Desertification is associated with previously arable land turned desert (one estimate is that approximately 251 000 sq. miles of farming and grazing land were swallowed up by the Sahara in the past 50 years) and its effects are often directly attributable to man's use or abuse of his environment (Eckholm & Brown 1977, p. 3). Environmental degradation is widespread and its effect on man's food and energy supplies became well known as the drought deepened in the Sahel of West Africa (Glantz 1977, pp. 1–15).

Drought, perceived as a climatic event, and desertification, defined as the spread of desert-like conditions due either to man or to climatic change, were initially discussed in highly technical terms (e.g. meteorological analysis of rainfall data and satellite imagery), which led to assessment of 'desert encroachment', 'ecosystem fragility', 'climatic modification', and the 'biogeophysical feedback mechanism'. The most protracted discussion concerned the temporal nature of the drought: was it cyclical and recurrent, or did it represent a permanent change in the world climate (Bryson 1973; Grove 1973, pp. 34–9; Nicholson 1980)? Less technical discussions focused on man's role in turning arable land to sand and its implications for mankind. Scholars considered the question 'The Sahel – does it have a future?' (Eckholm 1976) and, metaphorically, they asked 'Should the passengers on a rich lifeboat continue to admit castaways from crowded, poor lifeboats?' (Hardin 1974). Much evidence is marshalled to suggest that the honest answer to both questions is 'No'. Or, a qualified 'Yes' – providing the inhabitants drastically alter their life-styles and severely reduce their numbers. Otherwise, the future of the Sahel and of mankind is bleak.

On the other hand, although recurring drought is a normal part of Savanna–Sahel zones' climate (Nicholson 1979a, Baier 1980), the ecology of the Savanna–Sahel zones may be less fragile than previously thought and Sahelians more resilient. In 1974 the rains returned to near normal and, recently, scientists have reported that the long drought is ending, possibly by 1985 with full wet conditions being re-established around 1992 (Faure & Gac 1981, p. 477). However, Faure and Gac warn: 'If the same pattern continues, it is feared that a severe drought will occur around 2005' (Faure & Gac 1981, p. 477; see also Winstanley 1978). The Sahelians are also rebounding from the last drought, although scattered areas will continue to suffer from low rainfall.[2] For example, Niger has achieved grain self-sufficiency, exceeding its 1.5 million tons needed annually (Dash 1981). This grain self-sufficiency was accomplished, in part, 'from the government cutting taxes and the peasants then quickly turning over land that had been used for cash crops, such as cotton, to food crops' (Dash 1981). United States and other foreign assistance, as discussed below, have also contributed to this effort.

Still others argue, quite convincingly, that the devastation to the environment and the number of deaths in the human population may not have been as severe as first reported. For example, sober reflection and rigorous inquiry have revealed that the reported 100 000 deaths due to hunger in the Sahel between 1968 and around 1974 was sheer conjecture and may have grossly overstated the actual number of deaths (Simon 1981, p. 74). Simon goes further to indicate how inaccurate reporting may influence our approach to Sahelian development:

> To question the accuracy of the number of dead people in the Sahel may seem heartless, but the number is important: if it exaggerates the magnitude of the problem, we may despair of any solution. Undoubtedly a drought did occur in the Sahel, crops failed, people suffered, and some died. But to suggest that one crop failure was more severe than it was, and that another will be permanent, as the UN has done, is likely to convince us that there is no hope. (Simon 1981, p. 74)

All these studies contribute to our knowledge about drought, desertification, and about the Sahel region in general. Yet, however informative they are, they tended to overlook the politico-economic history of the region, the cultural attributes of the Sahelians, and the contribution Sahelians could make to the restoration of their own region. They seemed to put too much faith in technical rather than human or cultural solutions. They also placed great emphasis on Sahelians as victims of drought rather than of famine: of an event rather than a process. They blamed the victims for their despair, not their exploiters.

This book is more about famine than drought, more about the Savanna–Sahel zones than 'the Sahel' and more about human ecological adaptation than ecosystem abuse. It is also a book about rural economic development and the role of folk knowledge (practices) in land use and increased productivity. Two basic themes underlie the individual contributions in this book: (a) there is no creditable reason why local empirical experience (folk knowledge) and modern scientific understanding could not be combined to increase agricultural productivity in the Savanna–Sahel zones (Griffin 1978, p. 515); and (b) the basic cause of famine in the Savanna–Sahel zones is poverty. The poorest people starve because they have no money to buy food, they do not have other entitlements to exchange for available food, or because they have allocated their best land resources to nonfood crops (Lappe & Collins 1980, Sen 1981). This means that food self-sufficiency, a goal easily achieved by farmers and herders alike prior to European intervention, is now very difficult to achieve and is completely unobtainable during times of unusual environmental stress (drought). This loss of food self-sufficiency and the inability to purchase food have contributed to this region's increasing rural impoverishment. Analysis of how West Africans lost their ability to achieve self-sufficiency and their ability to purchase (or own) food goes a long way in helping us to understand the present problems posed by poverty in general and famine in particular.

Adjustments to the environment

According to Prothero, 'In physical conditions which are in all respects marginal, the relationships between people and land in cultivated or pastoral economies or some combination of these are very finely adjusted and delicately balanced' (Prothero 1974, p. 163). This has changed. The plight of the Savanna–Sahel zones must be seen as the result of an economic process, not merely an uncontrollable physical disaster. In the past, farmers and herders of the Savanna–Sahel zones had devised ways of adjusting to an environment in which seasonal and unusual droughts were common (Watts 1979, pp. 55–6). The meager, but growing historical record of this region indicates that neither farmers nor herders depended entirely upon the local resource endowment for their subsistence. Exchange, both between herders and farmers, and between regions to the north and south, complemented their own provisions (see Chs 2 & 3 and Baier 1980). Herders and farmers were extremely knowledgeable about their environment. Farmers, for example, distinguished microenvironments, several species of certain grains that they in turn associated with particular microenvironments (Ch. 3), and they had plant indicators to judge the conditions of soils (Haswell 1953, p. 32). Selected plants were also grown to stave off the inevitable hunger period (Haswell 1953, p. 25). During periods of unusually severe drought, farmers and herders would migrate to more bountiful areas where migrants engaged in wage employment or temporary cultivation of marketable crops. This kind of response to drought, called *cin rani* in Hausaland, is quite common throughout the Savanna–Sahel zones. But historically, migration has been seasonal or temporary. On the other hand, permanent emigration or migration is not characteristic of this area (Prothero 1974, pp. 166–7; Faulkingham & Thorbahn 1975, p. 470). The essential point is that these cultural adjustments even out food supply problems resulting from periodic food shortages, and droughts were far less destructive in the past than they are now (Baier 1980).

Herders devised specific strategies to cope with a variable environment. Some ultimately proved harmful to the ecology. One such strategy aimed at maximizing the number of cattle in their herd. The size of one's herd was a hedge against bad years (Swift 1973, p. 74), but it was also associated with good husbandry, individual productivity and wealth, and with personal prestige (Hopen 1958, pp. 23–8; Stenning 1959, p. 55). More importantly, large herds at maximum fertility are required to produce the greatest amount of milk, a basic staple in herders' diets (Stenning 1959, p. 102). Herders employ many coping strategies: herd diversification (Swift 1973, p. 73), long distance migration (Stenning 1959, pp. 206–25), a complex network of sharing, 'gifts', or loans among close relatives (Stenning 1959, p. 41; Swift 1973, p. 75), but the most effective 'risk-minimizing' strategy has proven to be 'holding' the maximum number of productive animals in the face of heavy losses (Baker 1974, p. 172). On the other hand, the size of the herd was limited by access to and the availability of drinking water, and by disease, especially rinderpest (de St Croix 1945, pp. 12–13). European intervention altered these limiting factors, but left the risk-minimizing strategy unchanged. The French introduced veterinary medicines, dug deep bore wells, and imposed political barriers to movement. Combined with a desire to maximize the size of herds, these technological and political factors resulted in a sharp increase in the number of animals (Wade 1974, p. 234) and in overgrazing of the accessible pastureland – a condition that was well established prior to the tragic drought of 1968–72 (for more on herders, see Ch. 8).

Increasing impoverishment

Food production in the Sahelian countries and in Africa generally has declined relative to that in other world regions (and relative to population growth). Besides the Near East region, Africa experienced the steepest decline in food self-sufficiency over the past 20 years (Norman 1981, p. 5).

There are many reasons for this decline. An obvious reason is the sharp reduction in the fallow period in many parts of Africa (Wade 1974, p. 236), resulting in soil depletion and reduced productivity. Low technology, insufficient land, and the displacement of small farmers by large, wealthy farmers acquiring farmland also placed limitations on food production (Haswell 1953, p. 59; see Chs 6 & 7). European penetration introduced several practices that ultimately reduced food production. The introduction of commercial agriculture was most destructive. The shift to cash-crop production meant that time (Haswell 1953, p. 59) and land (Anthony & Johnston 1968, p. 18) were diverted from staple crop production. In the Savanna–Sahel zones much of the best agricultural land was converted to the production of groundnuts (peanuts) and cotton, and the production of millet and sorghum declined.

In addition to land reallocation, social changes also impacted on food production. Understandably, wage-labor, either on European or African farms (Haswell, 1953, Anthony & Johnston 1968), reduced the time many farmers could devote to their own fields. But *gayya*, a form of communal labor traditionally employed to provide additional hands during certain bottleneck periods in staple crop cultivation, tended to break down with the production of commercial crops and under wage-labor (Anthony & Johnston 1968, p. 21). In conjunction with wage-labor and cash-crop production, which normally employed the most able-bodied young men, women were forced to assume full responsibility for food-crop production. This shift was particularly disruptive in the western part of the Savanna–Sahel zones where women normally supplemented the food supply by cultivating rice in river floodplains, called *fadamas* (Haswell 1953, p. 25).

More recent developments have also had the effect of limiting staple food production. Cattle manure, normally used as an organic fertilizer (Ch. 3), is now burned in parts of the region as fuel (Eckholm 1977, p. 11). If this practice should become widespread, it would have an even more devastating effect on staple food production. In addition, Haswell (1975) reported that through US foreign aid, grain was dumped in The Gambia, having the effect of reducing the economic incentives for farmers to produce their own grain. For all of the reasons given above, low production of staple foods cannot be directly related to soil quality, amount of moisture, or even length of growing season. Political and cultural factors also influence what is produced and by whom it will be consumed. Peter Taylor, a reporter for the British Broadcasting Corporation (BBC), lends credence to this view when he argues that 'millions of Malians are hungry not just because the soil is bad, the rains didn't come and harvest failed, [but] because they're poor; they're poor because they're powerless' (Taylor 1981, p. 1).

Another factor that contributes to the overall staple food shortage in the

Savanna–Sahel zones is the export of protein to developed countries, mainly in the form of fish. Recently, critics of US food trade policies have amassed a convincing set of data to demonstrate that, on balance, poor countries feed rich countries and that protein, especially fish, tends to flow from the more needy to less needy. George Kent made this point in *Development Forum*: 'One very clear illustration is provided by the fact that 56 million pounds of fish were exported from the famine stricken Sahel regions of Africa in 1971 alone' (Kent 1982). This case, though ironic, is not unique. Understandably, Kent concludes that 'One good way in which the U.S. [and other food importing nations] could help poor countries to increase their food self-sufficiency would be to increase its own food self-sufficiency' (Kent 1982).

There is another vital point that needs to be made regarding factors contributing to the decline of staple food self-sufficiency: that is, rapid population growth in farming communities, which forces the expansion of crop production onto ecologically unsuitable soils. Arable land and marginal land are surely being taken up, but not solely by farmers and not entirely in response to their growing numbers. The process is more complex, involving a near ancient process of Fulani sedentarization in towns, villages, and on land previously used for pasture (see Chs 3 & 4 and Frantz 1978, p. 103). The process was accelerated after European imposition of national boundaries, constraints on movements for tax purposes, and controlled ranching. Farmers' expansion into poor land was usually to grow groundnuts and cotton, while retaining as much as possible of their more fertile land to grow staple crops. At about the same time, especially during the decade of the 1950s, the Savanna–Sahel zones were experiencing a relatively wet period, but were moving rapidly toward a comparatively dry period of the 1960s, which reached its driest point during the 1968–74 drought (Nicholson 1979b, p. 623). New producers of groundnuts and cotton could not anticipate the impending decline in rainfall. Moreover, neither the French nor the British successfully introduced a technology to grow peanuts and cotton. Instead, they were cultivated without technological change (Anthony & Johnston 1968, p. 18). More serious than this, the man-made ecosystems so conducive for grain production were wholly unsuitable for groundnuts and cotton, resulting in their rapid destruction (Scott 1979). The end result, induced by both European policy and African land use, was the continuing breakdown of the Savanna–Sahel zones' ecosystem and a corresponding decline in staple food production.

Nigeria's drive for food self-reliance

Nigeria offers a specific case of a country currently experiencing severe food shortages that can be explained, at least in part, by agricultural policies originating during the colonial period and persisting up to the early 1980s. For this entire period, Nigeria has explicitly focused on irrigation farming to contribute to its national food requirements. Nigeria's experience is instructive in our continuing search for a humanitarian solution to the social and economic problems of the Savanna–Sahel zones.

Irrigation farming in the Savanna–Sahel zones and in Africa generally is surely an attractive solution to food scarcity because it raises the expectation of substantially increased agricultural production and of a reduction in the instability and uncertainty of farming on a rainfed basis. But these benefits, which come from the conversion of arid and semi-arid land into arable land, will depend on the provision of water through the control of large and small river systems. New technical inputs in agriculture will also require adequate and dependable supplies of water. According to *The Global 2000 report*, 'As increased pressure on supply generates wider use of high-productivity inputs, water management could become the single most important constraint on increasing yields in the developing world' (Council on Environmental Quality 1982, p. 100). Astonishingly, water management for irrigation farming accounts for only 3% of the cropped area in Africa (Council on Environmental Quality 1982, p. 151).

This latter fact is what makes irrigation farming so appealing. It means that the potential for irrigation cropping in Africa and in Nigeria specifically is immense (see Fig. 1.1). Although the actual extent of this resource is unknown, mainly because surveys have not been conducted in most African countries, the FAO (Food and Agriculture Organization of the United Nations) estimates that the Sahelian region of West Africa alone has about 12 million acres of land which is potentially available for irrigation if water is available (Eicher & Baker 1982, p. 134). The tantalizing feature of this estimate is that only 80 000 ha were under irrigation in the mid-1970s (Eicher & Baker 1982, p. 134). It is understandable, then, that Nigeria and other African countries are looking increasingly to large-scale irrigation farming projects to help meet the pressing demand for food.

Although Nigeria has emphasized irrigation agriculture for several decades, its severe food shortages are relatively recent (Oguntoyinbo 1970, Wallace 1980, Palmer-Jones 1981). Normally a food-exporting country, Nigeria is now a net importer of basic foods, including palm oil which is used locally in cooking and in industry. Since the mid-1970s, Nigerian food imports have exploded in amount and value. For example, 'From France alone Nigeria imported food worth $87 million in 1976, which [sic] figure jumped to $104 million [in] 1977' (Oculi 1979, p. 63). Overall, the total expenditure on imported foods rose rapidly from $297 million in 1975 to $439 million in 1976 (Forrest 1977, p. 78). Coupled with this increase in high value crops from abroad, the output of millet, sorghum, maize, and pulses actually declined. At that time, only rice and cocoyams showed an appreciable increase (Forrest 1977, p. 77). But by 1979, when President

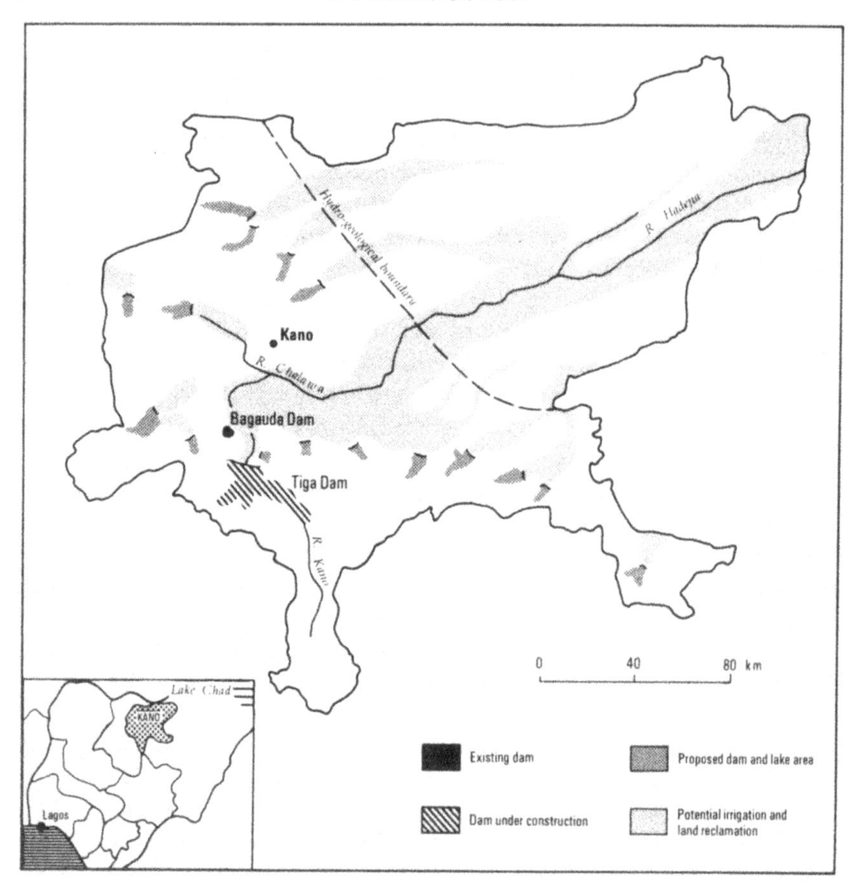

Figure 1.1 Potential water resources in Kano State. (*Source*: Kano Irrigation Project 1964.)

Shehu Shagari took office, Iroh reports that 'Nearly N15-million was being spent on rice imports alone, N4-million for meat; frozen and canned fish took another N145-million; maize wheat and wheat flour, N165-million, while milk and dairy products cost N135-million. These and other items amounted to over N1-billion, or 17 percent of the country's total import bill annually' (Iroh 1982, p. 45). Indeed, Nigeria had gone from the security of self-reliance in food to the costly situation of a net food importer.

Part of this increase in food cost was due to the high price of certain crops imported by Nigeria. Between 1962 and 1975 the combined price for millet, wheat, sugar, fish, and salt increased by 204% (Forrest 1977, p. 78). Not only was the price of certain commodities increasing very rapidly, but the quantity imported was also increasing. For example, 'the quantity of imported wheat increased at an annual compound rate of 23% (1962–75)' (Forrest 1977, p. 78). Another factor contributing to the near geometrical increase in the cost of food imports was Nigeria's shift to imports of prestige foods like wheat and rice, usually at high cost from exporting countries like the United States. At the moment, Nigeria is in the unusual position of being the second largest African producer of rice after Madagascar, but production has not kept pace with consumer demand which, since 1974, is increasing at an estimated rate of 13% per annum (Iroh 1982, p. 52). Rice, however, is not a staple in the diet and this recent demand comes from the rapidly growing urban population which can put pressure on the government to increase imports of this high valued commodity. Seemingly, to maintain political stability, the government will continue to import rice, but perhaps at a declining rate.

In order to meet the growing demand for rice and other basic foods, Operation Feed the Nation (OFN) was launched in 1976. OFN was only one in a relatively long series of government efforts beginning with the 3-year plan period of 1962–8 to increase food production by large- and small-scale farmers. A significant aspect of this series of efforts was the planned expenditure on irrigation farming, accounting for 5% of total expenditure in the 'First plan (N5 million), 11% in the Second plan (N24 million) and 26% in the Third plan (N425 million)' (Forrest 1977, p. 80). But these efforts failed. Now the civilian government of President Shagari has implemented its Green Revolution, the most comprehensive plan thus far to achieve food self-reliance, which is set for 1985.

The Green Revolution is not only the latest and most costly program yet – an unprecedented N89 billion – but it is also the most articulate (Iroh 1982, p. 48). Ambitious and expensive, the Green Revolution, like OFN, and to a lesser extent the plans that came before, focuses on both cash crops for export to encourage and revitalize the sagging palm oil, groundnut, cocoa, and rubber production, as well as on food crops for domestic consumption, especially nonstaple food crops to meet the demand generated in the urban sector.

The key features of the Green Revolution were critical aspects of OFN. Both cash and food crop production would be stimulated through increases in crop prices and other direct economic incentives to farmers. Research projects, improved agricultural services, and farm mechanization programs would support these efforts, mainly through planned service centers. Like OFN, the Green Revolution features a variety of farm subsidies, grants, and credit programs to promote fertilizer use and for farmers' cooperatives. In addition, OFN includes a concerted effort to produce protein-rich foods in the form of cattle, pork, poultry (for meat and eggs), and cow peas, a goal set by the Green Revolution as well (Iroh 1982, table I, p. 48).

As in other important aspects, OFN and the Green Revolution rely heavily on irrigation farming to help achieve food self-reliance. According to Iroh,

> The main superstructure of the Green Revolution programme is Nigeria's network of River Basin Development Authorities, designed to develop irrigation projects to fight drought in large parts of Nigeria, especially in the North. There are now no less than eleven such facilities, providing dams, farm water-works and irrigation canals on some 275,000 hectares of agricultural land. (Iroh 1982, p. 48)

Complementing these plans are the new Accelerated Development Projects, which are designed to provide several thousand kilometers of feeder roads, nearly 200 earth dams, and over 300 wells and boreholes. When completed, through an expected World Bank loan of $486 million, Accelerated Development Projects will involve all the 19 states of Nigeria (Iroh 1982, p. 48; World Bank 1981, pp. 5.44–5). Thousands of hectares of land will have been converted to irrigation production where rice, wheat, and other food crops will be grown for domestic consumption.

Although Nigeria has long looked to irrigation farming as an important, though partial, solution to its food problems (Abalu & D'Silva 1980), its record of accomplishments is not outstanding. Social critics of OFN and the Green Revolution, and especially of their reliance on large-scale irrigation projects, argue that large-scale irrigation projects will not sustain the hoped for 'green revolution' in Nigeria (Shantan & Watts 1979, p. 61) and that similar irrigation projects will not 'drought-proof' the Sahel either, a point discussed in detail below (Eicher & Baker 1982, p. 136). What these projects have done, they argue, is to disrupt the relationship between small-scale farmers and the land, and to benefit the large, more prosperous farmers, thereby exacerbating the already skewed income distribution. Some of the specific observations that lead them to this conclusion are discussed here. The three largest irrigation projects in northern Nigeria are the Kano River Project in Kano State, the Sokoto–Rima Basin Project in Sokoto and the South Chad Irrigation Project in Bornue State (Figure 1.1). They have also received the greatest criticism. The most general criticism is that large-scale irrigation projects do not perform as intended. Instead of improving the plight of the small-scale farmers through increased production and sales, and thus increased personal income, they are uprooted from their homes and land, often from valuable *fadama* land, and resettled on poor land with inadequate, incomplete, and often structurally unsound, housing (Wallace 1980, pp. 60–1). Farmers who lived down stream from the dams not only lost their rich *fadama* lands, but wells used to supply water for human and cattle consumption dried up and income from fishing declined in some areas (Wallace 1980, p. 65). Ironically, some projects created only slightly more irrigated land than existed as *fadama*. The Bakalori Dam, for example, had consumed N1 million by 1980 and flooded 20 000 ha of *fadama*, but only 1000 ha of newly irrigated land had been created (Wallace 1980, p. 65).

Critics have also questioned the massive expenditure on large-scale irrigation projects, given their modest record of success. Wallace reports that 'On the irrigation schemes wheat is yielding about 1.9 tons per hectare instead of the 4 predicted, and the sale of that wheat barely covers costs of production' (Wallace 1980, p. 69). Nevertheless, planned investment in large-scale irrigation farming continues. According to Forrest, 'An irrigated area of 274,000 hectares is proposed by 1991 at a cost of 220m naira at 1977 prices. These large projects are not based on any technical evaluation of the existing small and medium scale projects despite many statements of the need for such evaluation' (Forrest 1981, p. 241). Forrest goes further to question the basic economic rationale of the projects. Irrigation projects are justified as effective methods of producing wheat as a basic import substitute item, often amid unwarranted optimism. He observes that 'wheat imports have grown at a fast rate and it remains to be seen whether irrigated wheat can supplement and compete effectively with imported grain when low bread prices are important to those in power' (Forrest 1981, p. 241). Political constituents aside, it is simply more economical to import wheat than to grow it under irrigation. This has been known since at least 1966 when the FAO estimated that 'the cost of irrigated wheat production was $168 per ton as compared with a landed cost of imported wheat of $84 a ton' (Eicher & Baker 1982, p. 135). Finally, Watts questions the reliance on irrigation projects in arid regions because 'irrigation systems may, under extreme and recurrent drought conditions, become redundant due to water shortages which they were actually designed to prevent. Despite professions to the contrary, the inability of large-scale irrigation as a primary anti-drought strategy remains conjectural' (Watts 1977, p. 82).

Nigeria seems committed, mainly through inertia, to large-scale irrigation farming for the next two decades and the emphasis on wheat and rice production will no doubt continue, but the development of additional large-scale irrigation projects may be abandoned. Evaluation results from three World Bank funded Agricultural Development Projects in Gusau, Funtua, and Gombe areas of northern Nigeria indicate that 'compared to traditional plots, yields on fields cultivated with recommended practices and commercial input were more than twice as high. For sorghum and maize, results were particularly spectacular: improved local varieties reached 1,400 kilograms/hectare for sorghum and close to 2,000 kilograms/hectare for maize' (World Bank 1981, p. 5.44). Small-scale irrigation farming has proved viable. Indeed, 'The World Bank reports that Nigeria is shifting its emphasis away from large-scale irrigation to small-scale irrigation based on ground water development by hand operated and small motor-driven pumps' (Eicher & Baker 1982, p. 135). Ironically, small-scale irrigation farming of *fadama* lands using the shadoof or *jigc* (Hausa) to lift water to the surface is well established in Nigeria and other parts of the drought-stricken areas of the Sahelian region, including Sudan, Somalia, Kenya, and Tanzania, and could be the basis upon which a modern system of small-scale irrigation farming could be built. Obviously, much research is required. Still, small-scale irrigation farming should be seen as an important complement to rainfed farming if Nigeria's hope to achieve self-reliance in food production is to be realized.

Indices of poverty

The Savanna–Sahel zone countries, then, were deteriorating long before the drought of 1968–74. The drought accelerated the process to a crisis stage. Apart from oil-rich Nigeria which, as we have seen, imports much of its food requirements, all of the Savanna–Sahel zone countries and most of the 'drought-stricken' countries of East Africa are now deemed 'crisis countries' (Table 1.1). The term 'crisis countries' means that they are highly vulnerable economically or nutritionally, based on several indicators of development (Agricultural Development Indicators; International Agricultural Development Service 1980). They are among the poorest of all the underdeveloped countries and Table 1.1, which contains selected socio-economic indicators for 1981, suggests that they are in fact getting poorer.

Demographically, the Savanna–Sahel zone countries have a burgeoning population of 117 million, nearly 80 million in Nigeria alone (Table 1.1). The average growth rate for these countries is 2.6% per year.[3] This means that the populations of most of these countries will double in less than 25 years. Economically, the economies are heavily skewed, with on average well over 77% of the active population engaged in agriculture. They are essentially export economies, depending on the sale of one or a few primary commodities to generate well over 50% of their national revenues. Political and economic stability in these countries is highly vulnerable to fluctuations in world market prices, which are often low. The result is that the Gross National Product (GNP) per person for most of the Savanna–Sahel zone countries is uncommonly low. Furthermore, the growth in GNP between 1970 and 1978 was shockingly small. As Table 1.1 shows, during the period between 1970 and 1978 the GNP per person grew by less than 1% in five of the Savanna–Sahel countries. Apart from Cameroon, Mauritania, Nigeria, and Senegal, the GNP per person for all the Savanna–Sahel zone countries is below $300 per year, making these countries among the poorest in the world.

For most Savanna–Sahel zone countries, the staple crops are sorghum, millet, and maize (corn) (Table 1.1). As expected, Table 1.1 shows that per capita consumption of grain is very high, ranging on average between 122 kg per year in Cameroon and 319 kg per year in Niger from 1973 to 1979. Overall, grain yield per hectare is relatively low and average per capita food production during 1969–71 versus 1973–79 has declined by 1.2% to about 1.3% in 1980, a rate considerably lower than the rate of population growth. In every Savanna–Sahel zone country, and in most of the drought-stricken ones of East Africa, essential grains must now be imported, requiring the expenditure of needed foreign exchange. Cape Verde and Mauritania are extreme cases, importing over 80% of their total grain consumption between 1977 and 1979, but Senegal and The Gambia imports are surprisingly high, since these countries were food self-sufficient in the recent past (see Haswell 1975, above).

On the other hand, their current dependency on imported grain may be related to production priorities. For example, Senegal and The Gambia allocate an uncommonly high percentage of their total arable land to cash crops: 40% to staples : 55% to cash and 15% to staples : 73% to cash, respectively (Table 1.2). This is consistent with the trend in West Africa generally, where the production of export crops has increased markedly over the past decade (Shear & Stacy 1975, p. 9), while food production has declined. Another striking observation when all of the countries are considered is that the percentage of arable land actually in cultivation is surprisingly low, which is an indication of the enormous agricultural potential of this region and of Africa generally.

Nevertheless, food self-sufficiency has dropped and the quality of life has worsened. For example, Table 1.2 shows that in 1973 all of the Savanna–Sahel zone countries had a deficit in caloric supply relative to requirements. The largest deficits were in Upper Volta, Chad, Mauritania, Niger, and Nigeria, whose supplies of calories were between 15 and 30% below the minimum requirements. For Senegal and The Gambia, the deficits were less than 6%. These data highlight the importance of grains in the diet. Comparable data on protein deficiency are not available, but the importance of fish, and to some extent beef and poultry, in the regional diet is well known. In addition, Table 1.2 shows that life expectancy at birth (1970–75) in this region is relatively low (average 39 years), with Senegal, The Gambia and Nigeria among the highest.

Clearly, neither the drought of 1968–74 nor the 'encroaching desert' caused the current level of impoverishment in the Savanna–Sahel zones.[4] The poverty we now observe and the famine this region experienced are the result of a process involving the imposition of a European, extractive economy, the rapid growth in both the cattle and the human populations, the increasing sedentarization of Fulani nomads, and the establishment of a system of low wage labor, burdensome taxes, and forced labor, which limited or took away entirely the ability of West Africans to obtain food.

Table 1.1 Selected socio-economic indicators, 1981[a]

	West African Savanna–Sahel countries				
	Cape Verde	Nigeria	Chad	Cameroon	Mauritania
demographic					
crisis countries					
population in 1980 (millions)	0.3	77	4.5	8.5	1.6
rate of natural increase in 1978 (% per year)	1.8	3.2	2.3	2.3	2.8
population in agriculture in 1979 (%)	57	54	84	81	83
politico-economic					
leading exports, 1976[b]	NA	Pet, C, Pe	Cot	C, Co, A	I
percent 1971–3 export, 1976[b]	NA	88	66	54	79
GNP per person (at market prices) in 1979 (US$)	270	670	110	560	320
GNP growth, 1970–8[c] (% per year)	NA	4.4	−0.6	2.8	−0.6
agricultural production					
major cereal crops (1977–9)	M	S, M	S, R	S, M	S
cereal output, avg. 1977–9 (million tons)	4 thou.	9.2	0.6	0.9	30 thou.
cereal area, avg. 1977–9 (million ha)	10 thou.	13	1.2	1.0	0.1
cereal yield, avg. 1977–9 (kg ha^{-1})	400	700	500	900	300
change in cereal output (% per year^{-1} 1969–71 vs 1977–9)	9.1	1.6	−0.9	2.8	−11.3
change in cereal area (% per year^{-1} 1969–71 vs 1977–9)	12.1	0.8	2.4	2.9	11.3
change in cereal yield (% per year^{-1} 1969–71 vs 1977–9)	−2.0	0.8	−3.2	−0.1	0.5
food consumption					
cereal use per person, avg. 1977–9 (kg year^{-1})	212	157	153	126	122
import content in cereal use, avg. 1977–9 (%)	94	15	3	12	82
agricultural inputs					
economic land in crops, 1978 (%)	61	32	30	16	<0.5
cropped land irrigated, 1978 (%)	5	<0.5	<0.5	<0.5	4
cropped land per person, 1978 (ha)	0.1	0.4	0.5	0.9	0.1
fertilizer use, 1978 (kg ha^{-1})	3	3	4	5	10
land area by zones (%)					
Sahara	NA	NA	50	NA	77
Sahelian	NA	NA	21	NA	23

NA = not available.
M = maize; W = wheat; R = rice; S = sorghum and millet; C = cocoa; Co = coffee; A = aluminum; D = diamonds; U = uranium; P = phosphates; Pe = peanuts; L = livestock; I = iron ore; Cop = copper; H&S = hides and skins; T = tea; Pet = petroleum products; Cot = cotton.

West African Savanna–Sahel countries					Selected East African drought-stricken countries			
Mali	Niger	Senegal	The Gambia	Upper Volta	Kenya	Ethiopia	Tanzania	Zambia
6.6	5.5	5.7	0.6	6.9	16	33	19	5.8
2.7	2.9	2.6	2.4	2.6	3.9	2.5	3.1	3.2
87	89	75	78	82	78	80	82	67
Co, L	Pe, U	Pe, P	Pe	L, Cot	Cot, Pet	Co, H&S	Co, Cot, D	Cop
64	60	40	98	59	38	57	37	93
140	270	430	260	180	380	130	270	510
1.8	−0.6	−0.3	2.9	−1.0	2.5	−0.1	1.7	−0.9
S, R, M	S	S, R	S, R	S, M	M, W	S, W, M	M, S, R	S, M
1.2	1.6	0.7	0.1	1.1	2.8	4.1	1.6	0.9
1.7	3.6	1.0	0.1	2.1	2.0	4.6	2.0	1.2
700	400	700	800	500	1400	900	800	800
1.4	2.5	0.5	−4.3	1.8	0.4	−0.6	2.6	−0.1
4.3	2.5	−1.2	−1.8	0.4	1.0	1.0	2.4	−0.1
−2.7	0	1.7	−2.6	1.4	−0.7	−0.7	0.1	−0.1
191	319	210	182	184	186	142	103	187
3	2	36	40	5	0	3	3	12
6	13	18	40	25	29	16	6	7
5	1	5	10	<0.5	2	<0.5	1	<0.5
0.3	0.6	0.4	0.4	0.9	0.2	0.5	0.3	0.9
8	1	13	16	2	23	2	16	14
50	65	50	NA	0.0	NA	NA	NA	NA
25	30	37	NA	13	NA	NA	NA	NA

Sources:
[a]International Agricultural Development Service (1980, pp. 14–15).
[b]Agency for International Development (1976).
[c]World Bank (1980, pp. 6–7).

Over time, Savanna–Sahelians lost entitlement to food, specifically through an inability to produce it and to purchase it. This process of impoverishment exacerbated the effects of the drought. This means that the Savanna–Sahelians are victims of famine, not drought.

Reclaiming the land

Restoration of the land in the Savanna–Sahel zones will require much time and money. For one thing, the region is large (2.044 million sq. miles) and both drought and famine are spreading to countries that once exported food, but are now importers (see Fig. 1.2 and Table 1.1). (We should be reminded that famine in East Africa, especially in northern Uganda and Kenya, Ethiopia and Somalia, and Zambia and Zimbabwe, though they are

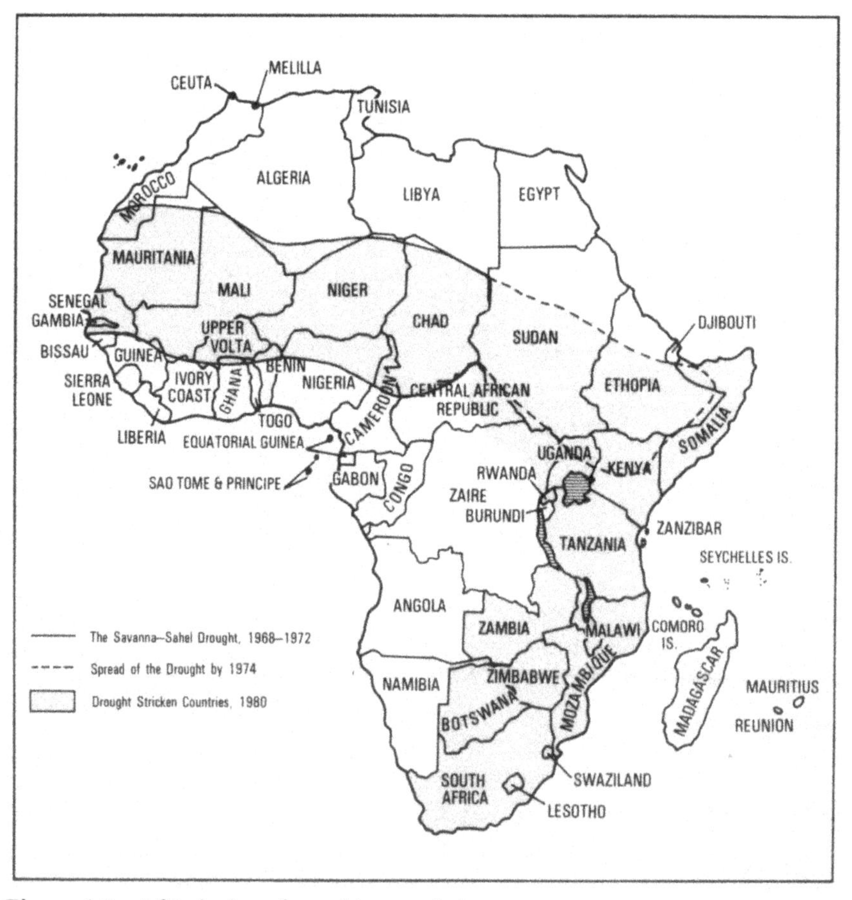

Figure 1.2 Africa's drought and hunger belt.

drought stricken, is due mainly to war and internal conflicts. In addition, increasing population, declining food production, government indifference, and greed have all contributed to the famine (Proffitt 1980, p. 48). Furthermore, the Sahelian countries are landlocked and movements between the coast and the interior and from one location to another within the region are difficult and costly (Simon 1981, p. 74). Consequently, the initial costs and the 'recurrent costs' of reclaiming the land will exceed the combined GNP of the Sahelian countries (W. Derman, personal communication).

The money needed to defray the cost of the initial recovery projects will be raised by the newly formed Club du Sahel and the Permanent Interstate Committee for Drought Control in the Sahel (The Club–CILSS). To date they have been successful. According to Fell, 'Concessional commitments to the Sahel rose from $755 million in 1975 to $1.7 billion in 1979, more than a 50% increase in constant terms and more than three times the per capita level of increase registered by the rest of Africa' (Fell 1981, p. 18). But the 'recurrent costs' (i.e. the additional cost of building maintenance, operational costs, etc.) are also high and must be assumed entirely by the Sahelian countries. Project analysts discovered 'that in 1982 the countries of the Sahel would show a deficit of $180 million, [which is] the difference between expenditure commitments for public programs and planned receipts' (Fell 1981, p. 19). If this deficit is not made up, recovery projects will suffer. In any event, the countries will suffer for years to come because they will need to tax their populations more, reduce government jobs and services, and undertake other austere measures to generate the money, or they must become even more dependent on uncertain and politically weakening foreign aid, undoubtedly in the form of equipment, material, and skilled advisers.

The United States is an important member of the donor countries, who are contributing to efforts to reclaim the land, allocating $107.5 million for the Sahel (McPherson 1981, p. 5) or approximately 8% of the total in an average year (Fell 1981, p. 20). United States donations are part of an overall commitment to the development of the Sahel. After the Sahelian drought peaked around 1974, Secretary of State Henry Kissinger, in a keynote address to the World Food Conference, announced that 'today we must proclaim a bold objective – that within a decade no child will go to bed hungry, that no family will fear for its next day's bread, and that no human beings' future and capacities will be stunted by malnutrition' (Kissinger 1974, p. 1). A year and a half later, on May 1, 1976, in an address delivered in Dakar, Senegal, Secretary of State Kissinger called for 'a comprehensive intervention program that, rather than ease the effects of the drought, will help rollback the desert; instead of relief measures, will develop additional water resources, increase crop acreage, and build food storage facilities in order to ensure that hard-won programs in economic development will not continue to be wiped out by recurrent natural disaster' (Kissinger 1976, p. 11). With these proclamations, the United States adopted the Sahel as its world region for development.[5]

To 'rollback the desert', many recovery projects, usually large scale and on a regional basis, were identified and implemented. Shear and Stacy recognized the scope of the undertaking:

> Generally, any sound livestock strategy or even individual project design must consider the inter-relationship between the Sahelian, Soudanian, and Guinean Zones, as well as the commercial inter-dependence between coastal and interior states. The special potentialities and resources of each zone should be maximized in a way that does not degrade the environment. (Shear & Stacy 1975, p. 10)

Although the need for a regional approach was clearly recognized, most of the projects 'targeted an area or an ethnic group, often the nomads' (see Ch. 8). The targeted areas usually involved large-scale river basin control to expand the amount of land under 'controlled irrigation' (Worthington 1981, p. 16). As we have seen, controlled irrigation is perceived as the key to the recovery of the Sahel; that is, to achieve food self-reliance and sustained economic growth without environmental damage by the year 2000. As was done in Nigeria, the region's large river basins will be used (see Ch. 7): these include the Senegal, Niger, Sokoto, and Kano Rivers and many other minor streams.[6] For example, multipurpose dams are scheduled for construction at Manatali and Diana to prevent salt-water intrusion, to improve navigation, and to open up nearly 800 000 acres for irrigated agriculture (Worthington 1981, p. 17). Initial reports from Bakel,[7] an area being developed with USAID (Agency for International Development) and Senegalese funds, are impressive, indicating that much land is now productive, food production has increased, marketable surpluses are now produced, environmental risks from variable rainfall and pests are reduced and, more importantly, this record of success has been achieved with farmers' involvement.

However successful these river basin control projects appear to be they, like the large irrigation projects in Nigeria, have been severely criticized. The most comprehensive critique of the Bakel Project and many other recovery projects is found in *Seeds of famine* by Franke and Chasin (1980). Although Worthington stresses the involvement of farmers through their 'groupments' or production units for the initial success of the Bakel Project (Worthington 1981, p. 17), Franke and Chasin argue that 'USAID has effectively joined SAED (Société d'Aménagement et d'Exploitation des Terres du Delta du fleuye Sénégal) in a political campaign to thwart local initiative and impose a development model incompatible with local social and ecological conditions' (Franke & Chasin 1980, pp. 219–20). Specifically, the authors criticize SAED's efforts, in collaboration with USAID, to force the production of rice for the urban market, requiring expensive inputs and inducing severe indebtedness, as opposed to the production of millet and sorghum, the local staples, using inexpensive tools and organic fertilizer acquired locally. They also argue that SAED and USAID, in addition to causing 'harmful economic effects' by insisting on rice production, are making a concerted effort to break up the original federation of farmers by demanding that farms be held in individual tenure rather than collectively, by taking control of the leadership positions in the federation, by displaying a complete insensitivity to potential health hazards from malaria and schistosomiasis, and by transforming the project into a profit-making operation for SAED bureaucrats and their Sahelian representatives (Franke & Chasin 1980, pp. 221–6). These efforts will not only lead to the destruction of the federation and increasing commercialization of rice production, but may also lead to a reduction in staple food production and further destruction of the environment.

The United States is involved in other recovery projects where good intentions are lost in a quagmire of local government policy, farmers' indifference, and poor planning. Together, they could have a devastating effect on food production. Journalist Peter Taylor of the BBC reports from Mali, where the United States is financing a $10 million project – Operation Millet – to help the country achieve self-reliance in food, that:

> They've spent a third of the $10 million on building roads. The biggest part has been spent on building warehouses, homes for the officials, offices, and in paying general administrative costs. Only a small part has been used to aid the peasants directly – a credit fund to help them buy the plows and oxen which to them represent advanced technology.
>
> [Operation Millet is designed to achieve food self-sufficiency by increasing the production of millet], but the Americans have found many obstacles in their path. The 65,000 farmers in the Operation Mils area are deeply cautious and loathe to take risks. They're reluctant to borrow money to buy the new equipment and fertilizer they need to increase production. [Even if production is increased, the surplus may not be distributed to other parts of the country.] In a good year, the farmer's instinct is to hoard his grain, an insurance for the bad years to come. Without these reserves, farmers know that when the harvest fails, they'll have to pay dearly for the millet they need to keep their families alive. But of course they have to pay taxes and buy necessities so some of their millet is sold at market. But even then it doesn't benefit the regions of Mali which are short of food. The markets are across the border. Here, the price of millet is twice that in Mali. Most of Mali's millet and sorghum is grown in the regions around Mopti and the south of the country. From here much of it – an estimated 20 percent – is smuggled across the border into Upper Volta. There it may

> be sold within the country or be passed on to Niger, Nigeria, and Ghana where prices are still higher. A similar leakage of grain goes on across all of Mali's long borders into the Ivory Coast, Guinea, Senegal, and Mauritania. This is why, even in a good year, peasants who live in Mali's deficit regions do not have enough to eat.

Taylor goes further to show how a government agency, with US reluctant duplicity, exploits the farmer to appease urban consumers, who are more able to apply pressure on the government.

> The Office du Produits Agricoles du Mali – OPAM – is the main reason for the smuggling. OPAM is the instrument the government uses to extract cheap food from the peasants to feed the army, the police and the bureaucracy. OPAM has a legal monopoly. Officially, the peasants are not allowed to sell their grain to anyone else. The price OPAM offers is so low that without that monopoly it would acquire no grain at all. In theory, OPAM's purpose is to supply the town and regions short of food, but few of the sacks in these lorries will ever leave Bamako. This is how OPAM gets its grain. The process is called 'commercialization,' and the peasants hate it . . . the peasants have no choice [but to sell millet at fixed prices and quotas to OPAM.] . . . USAID found itself caught in the middle. The government decided that in the Mopti area, Operation Mils should collect the fixed quotas from the peasants. USAID provided the transport and the personnel. The result was that the peasants – who were supposed to look upon Operation Mils as a friend – now equated it with the hated OPAM. USAID recognizes that the last people whom its $10 million project helps are the poor farmers themselves.

In summary, Taylor reports that the United States

> admits that only 3 per cent of all farming families receive direct benefits, and that all families in this area bear the burden of being forced to sell their fixed quotas to the state. [But] whatever its [Operation Millet] shortcomings, it's only this kind of aid which can help Mali escape the grip of poverty and hunger. (Taylor 1981, pp. 4–5)

Other critics argue that river basin control would disrupt life in the region and that, technically, irrigation agriculture is not a wise investment of development capital in semi-arid regions. In the first instance, river basin projects would prevent the traditional migration movement and grazing of river floodplains by Fulani and other nomads, and they would prevent *fadama* cultivation, an exceedingly important source of dry season food. Nomads and farmers would be forced into new life-styles involving unfamiliar technology and possibly imposed organizational structures (Derman 1981). There could be further erosion of their control over their own resources. In the second instance, regional irrigation programs may be poor investments because: (a) available evidence shows that the return on existing schemes is low, (b) low rates (cost) essentially amount to a subsidy to a minority of usually better-off farmers (unreasonably low rates also lead to waste through overwatering), (c) water from large-scale projects is often underutilized, and (d) the financial return to the government is low (Kumar 1977, pp. 481–2). This last point seems to be especially true for the Savanna–Sahel zones, where local governments, rather than donor countries, must assume the disproportionately high 'recurrent costs' on recovery projects, which will continue to absorb their development capital (Derman 1981).[8]

There is no right answer or ultimate development strategy for the Savanna–Sahel zones. Instead, several strategies must be employed. More importantly, a compromise must be found between the old and the new, between existing land-use patterns and introduced (proposed) ones, between labor and capital intensive projects, between staple food for local consumption and commercial crops for urban and/or overseas markets, between local initiative and external control. More importantly, if the Nigerian experience with large-scale irrigation farming is a useful guide for other Savanna–Sahel zone countries, small-scale irrigation projects, hopefully based on farmers' knowledge of *fadama* cultivation, should be given serious consideration. Because we feel so strongly on this point and on the underlying philosophy presented above, the approach taken here is to discuss 'life before the drought', which offers information on social structure, ethnic interdependency, ethnoscience (i.e. farmers' empirical and cognitive knowledge of their environment which is used in agricultural decision-making), and river basin management that must be considered in any development program for Savanna–Sahel zones. Now that the flurry of writing (on 'the drought') has waned and the efforts to lay blame have declined, it is time to discuss the people of this region and their long-standing methods of coping with their environment.

Outline of the book

In Chapter 2 the social history of the Songhay Kingdom between 1464 and 1591 is presented as an explanation of the spread of Islam, the mode of production, and the institution of slavery. There is much confusion surrounding the repressive and progressive aspects of Islam regarding socio-economic development and the nature of slavery in West Africa generally. Too often Islam is seen as a barrier to economic progress and slavery as too benevolent. Employing primary sources extensively, these topics are analyzed and clearly put into historical perspective. The mode of agricultural production, including land ownership, the organization of labor, and the distribution of surplus, are also analyzed and similarities between the 'past' and 'present' are highlighted.

Social scientists and journalists often stress ethnic differences between herders and farmers in the Savanna–Sahel zones and intergroup hostilities and conflicts, which are associated with access to land and water, as fundamental to inter-ethnic interaction in the Savanna–Sahel zones. Chapter 3 rejects, at least partially, the notion that herders and nomads are in perpetual conflict and illustrates the well established areas of mutual cooperation and interdependence. The approach is human ecological and shows that one response to regular environmental stress is intergroup cooperation, which enhances the survivability of both groups.

Chapter 4 explores the paradoxical relation between the established role of the state in the development of mechanized, irrigation farming, which is described here as a highly extractive (exploitive) pattern of agricultural development, and the growing pressure of land scarcity faced by the agricultural and pastoral peasantry. The irony developed in this analysis is that of the relative food security of the Sudan, even during severe drought such as the one of 1968–74, which is due to the 'success' of large-scale, mechanized irrigation farming while the masses of the farm peasants and pastoralists are deprived access to land and other national resources, principally water. The dilemma is this: the state is now dependent on this extractive economy, well-to-do merchants are investing more of their profits, along with subsidized loans from the state, in mechanized, irrigation agriculture, and the peasantry, either due to poverty or lack of collective will, are unable to invest in an alternative structure. The result is that the old land-use patterns and the social relations, those established prior to independence in 1956, persist. Drought, then, is not seen as a stimulus of change, but as an omen to expand the established land-use pattern and intensify the old social relations.

Chapter 5 shows how farmers cope in a harsh environment, mainly through mutual interaction with other ethnic groups, and how European intervention disrupted the livelihood systems that were so suitable for given parts of Africa. The Machakos District is geographically distant, but it is uncommonly similar to regions like Hausaland of West Africa regarding farmers' adaptive strategies and paths to development.

Famine in the Savanna–Sahel zones is said to have been caused by the 1968–74 drought. Chapter 6 offers evidence to the contrary. Instead, British penetration, the imposition of a cash economy, and the allocation of arable land to peanut production resulted in farmers losing control over their land and losing their ability to achieve self-sufficiency. These are symptoms of increasing impoverishment in the Savanna–Sahel zones and the available historical evidence suggests that they are the result of a process, one that was accelerated after British intervention and maintained after independence. The links between drought and famine, and between social and political structures and famine, are made clear. Poverty had reached desperate proportions prior to the drought.

Resource endowment of the Savanna–Sahel zones is often believed to include its subsurface mineral wealth, its arable land, and its grazing lands. Another resource that is well known, but seldom discussed in detail, is *fadama* land. In Chapter 7 *fadama* (pl. *fadamas*) land is discussed as a single land type or resource. Though small in total area relative to upland in the Savanna–Sahel zone, *fadama* is far more productive (per unit of cultivation), presents less production risk, requires higher levels of technological skill, and offers the greatest potential for agricultural development. Over time, various levels of technology have been employed to exploit *fadamas*, some with greater success than others. These levels of technology and their relative levels of exploitive efficiency, from the point of underutilization to technology overload, are demonstrated in a detailed study of *fadamas* near the ancient city of Zaria, North Central State, Nigeria. The agricultural, and now financial, value of *fadamas* is emphasized and their potential role in the Savanna–Sahel recovery is clearly stated.

Finally, the nomads, held in utter contempt prior to the drought by local governments and donor countries alike (Baker 1974, pp. 171–2), are now perceived, along with their cattle, as the major victims of the drought and the brunt of resentment and abuse by local farmers. Because of this new perception, mainly by external agencies, many recovery projects have focused on the needs of the nomads. They are now the 'targeted' group. In Chapter 8 the relationship of the Fulani to their cattle and their environment is discussed to provide background for USAID decisions regarding how its assistance should be administered. Social scientists, especially anthropologists, are usually skeptical of foreign intervention into a non-Western culture. Though well intended, foreign aid seldom reaches the 'poorest of the poor', and the projects are often insensitive and disruptive. To counteract much that has already been done and to anticipate the effects of proposed recovery projects, Chapter 8 offers a prescription for development. Significantly, the prescription is not a cure for 'desertification', said to be caused by man's irrational actions, or simply one for 'overgrazing', by herds too large to be sustained on the meager grass cover. Instead, the prescription calls for compassion and understanding.

Let me focus your attention on one final point. Although none of the chapters offer political recommendations, apart from Chapter 8, they all address the issue of rural economic development and suggest alternative courses of action. The reader is urged to evaluate the history and nature of the relationship between ethnic groups, West Africans' folk knowledge of their environment, and various adaptive strategies for an uncertain and harsh environment that could be combined with modern technology, new land-use practices, and new marketing structures to improve overall agricultural productivity without environmental degradation.

Notes

1 The Sahel countries include Cape Verde, Mauritania, Senegal, The Gambia, Upper Volta, Niger, and Chad. The Savanna–Sahel zones include, in addition to the Sahelian countries, the northern part of Nigeria and Cameroon. In this introduction, Kenya, Ethiopia, Tanzania, and Zambia are included for comparative purposes.

2 According to James Delehanty (personal communication), Niger was very proud of achieving self-sufficiency in grain in 1980, although Bande, south of Zinder, and other isolated regions suffered from rainfall deficits and the prospects for continued self-sufficiency looked much less promising in mid-July, 1980. Also, Niger's groundnut crop was struck by a near fatal peanut blight in 1974 that has persisted to the present.

3 The Sahel and East African countries have a population of 191 million. Apart from Cape Verde Island, the average growth rate is 2.9% per year, among the highest in the world. Kenya has the world's highest annual population growth rate of 3.9%.

4 Recently, a great deal of evidence has been offered to show how multinational agribusinesses are making the Savanna–Sahel zones produce, but not for the domestic market. Instead, the huge volume of food produced by these firms go to Europe and North America where consumers are able to pay premium prices. The 'successful' operation of multinational agribusinesses in a harsh environment has led scholars to argue, as we have, that 'Drought is a natural phenomenon. Famine is a human phenomenon' (see Lappe & Collins 1980, pp. 94–5 & 286–8). The same argument is made by deSouza and Foust (1979, 498–501).

5 The Sahel, though long designated by Arabic travelers and scholars as the 'edge' of the desert, is an unusual 'region'. It evokes more emotion than geography, economics or history. The cultural and ethnic reality of the region is lost in a maze of concern for man's abuse of his environment, man's exploitation of man, and man's potential for destroying others while he kills himself. The Sahel seems so cut and dry; one can easily 'identify' the Sahel (or identify with it). Blame can be laid squarely (at the feet of the colonial French, the romantic nomads, or the industrial West). More important, humanitarian aid can be focused on it. Recovery could be monitored and clearly cataloged.

6 Controlled irrigation is a well established strategy for reducing environmental risks in the Savanna–Sahel zones. Many projects have been disappointing and subsequently abandoned. Though Senegal, Niger, and Mali stand out as Sahelian countries (The Gambia is an exceptional case) where controlled irrigation contributes substantially to overall food production (see Table 1.1), Nigeria is also very heavily committed to irrigation farming (see, for example, Oguntoyinbo 1970 and Ch. 7).

7 The Bakel Project, located on the upper valley of the Senegal River on the border between Mauritania and Senegal, is an important test case for irrigated farming techniques under the auspices of USAID and SAED, the Senegal government's Society for the Development and Exploitation of the Land and the Delta (along the Senegal River for rice production). Other donor countries are monitoring this project and their participation in subsequent collaborative efforts will surely be based on the success or failure of large-scale irrigation farming on the Bakel Project.

8 The fear, expressed by W. Derman (personal communication), is that the Sahelian countries will become politically and economically dependent on Western donor nations, especially the United States, and Sahelian governments will thereby lose control over future development in their own countries. Marilyn Clement voiced a similar warning when she

said, 'Whatever aid our [US] government gives to help alleviate this problem must not be designed to perpetuate economic dependency or political conformity in the recipients' (Clement 1974).

References

Abalu, G. O. I. and B. D'Silva 1980. Nigeria's food situation – problems and prospects. *Food Policy* **5**, 49–60.

Agency for International Development 1976. *Africa – economic growth trends*. Statistics and Reports Division, Office of Financial Management, Bureau for Program and Management Services.

Agency for International Development 1980. *Africa*. Congressional Presentation, Annex I.

Anthony, K. R. M. and B. F. Johnston 1968. *Economic, cultural and technical determinants of agricultural change in tropical Africa. Field study of agricultural change, northern Katsina, Nigeria*. Prelim. Rep., no. 6. Food Research Institute, Stanford University.

Baier, S. 1980. *An economic history of Central Niger*. New York: Oxford University Press.

Baker, R. 1974. Famine: the cost of development? *Ecologist* **4**, 170–5.

Bryson, R. 1973. Drought in Sahelia: who or what is to blame? *Ecologist* **3**, 366–71.

Clement, M. 1974. World hunger crisis escalates. *Interreligions Foundation for Community Organization News* **5**, 4.

Council on Environmental Quality 1982. *The Global 2000 report to the President*, vol. 1. Council on Environmental Quality and the Department of State. Washington, DC: Penguin.

Dash, L. 1981. Struggling Sahel region farmers learning to use oxen, fertilizer. *Washington Post*, April 14.

Derman, W. 1981. *The Sahelian drought and famine reconsidered*. Paper presented at the Spring Symposium on Africa in the Eighties, University of Minnesota, May 6–8, 1981. Presentation taped by the editor.

Eckholm, E. 1976. The Sahel – does it have a future? *Africa Rep*. **21**, 12–16.

Eckholm, E. 1977. The other energy crisis: Finewood. *Focus* **XXVII**(4), March–April, 9–16.

Eckhom, E. and L. R. Brown 1977. The spreading deserts. *Focus* **28**, 1–11.

Eicher, C. K. and D. C. Baker 1982. *Research on agricultural development in sub-Saharan Africa: a critical survey*. MSU Int. Develop. Pap., no. 1. Department of Agricultural Economics, Michigan State University, East Lansing, Michigan.

Faulkingham, R. and P. Thorbahn 1975. Population dynamics and droughts: A village in Niger. *Popul. Stud*. **29**, 463–77.

Faure, H. and J.-Y. Gac 1981. Will the Sahelian drought end in 1985? *Nature* **291**, 475–8.

Fell, A. M. 1981. The high cost of recovery in the Sahel. *Agenda, Agency for International Development* **4**, 18–20.

Forrest, T. G. 1977. The economic context of Operation Feed the Nation. *Savanna* **6**, 77–80.

Forrest, T. 1981. Agricultural policies in Nigeria, 1900–78. In *Rural development in tropical Africa*, J. Heyer, P. Roberts and G. Williams (eds). New York: St Martin's Press.

Franke, R. W. and B. H. Chasin 1980. *Seeds of famine – ecological destruction and the development dilemma in the West African Sahel*. Montclair, NJ: Allanheld, Osmun.

Frantz, C. 1978. Ecology and social organization among Nigerian Fulbe (Fulani). In *The nomadic alternative*, W. Weissleder (ed.). The Hague: Mouton.

Glantz, M. H. 1977. The U.N. and desertification: dealing with a global problem. In *Desertification – environmental degradation in and around arid lands*, M. H. Glantz (ed.). Boulder, Colorado: Westview Press.

Griffin, J. 1978. World agriculture: the need for a scientific base. *New Scient*. **80**, 514–16.

Grove, A. 1973. Desertification in the African environment. In *Drought in Africa*, D. Dalby and R. Church (eds). Report for the 1973 Symposium, Center for African Studies, University of London.

Hardin, G. 1974. Living on a lifeboat. *Bioscience* **24**, 561–8.

Haswell, M. R. 1953. *Economics of agriculture in a savanna village*. London: Her Majesty's Stationery Office, for the Colonial Office.

Haswell, M. R. 1975. *The nature of poverty*. London: Macmillan.

Hopen, C. 1958. *The pastoral Fulby family in Gwandu*. Oxford: Oxford University Press.

International Agricultural Development Service 1980. *Agricultural Development Indicators – a statistical handbook*. New York.

Iroh, E. 1982. Nigerian agriculture: how the rush for petro-naira put the skids under agriculture. *New African*, no. 179, 45–63.

Kano Irrigation Project 1964. Husbanding water supplies for the benefit of Nigerian agriculture and industry. *Stand. Bank Rev*. November, 3.

Kent, G. 1982. The poor feed the rich . . . *Develop. Forum* **10**, 5.

Kissinger, H., Secretary of State 1974. Keynote address to World Food Conference on November 5, 1974. *War on Hunger* **8**, 1–4; 21–4.

Kissinger, H., Secretary of State 1976. New U.S. initiative on Sahel. *War on Hunger* **10**, 10–13.

Kumar, D. 1977. The edge of the desert: the problem of poor and semi-arid lands. *Trans. R. Soc. Lond*. **278**, 477–91.

Lappe, F. M. and J. Collins 1980. *Food first – beyond the myth of scarcity*. New York: Ballantine.

McPherson, P. M. 1981. *Development assistance for the Third World*, 1–6. United States Department of State, Bureau of Public Affairs, Washington, DC.

Nicholson, S. E. 1979a. The method of historical climate reconstruction and its application to Africa. *J. Afr. Hist. Vol*. **20**, 31–49.

Nicholson, S. E. 1979b. Revised rainfall series for the West African subtropics. *Mnthly Weather Rev*. **107**, 620–3.

Nicholson, S. E. 1980. The nature of rainfall fluctuations in subtropical West Africa. *Mnthly Weather Rev*. **108**, 473–87.

Norman, D. 1981. Progress or catastrophe in Africa? *Africa Report* **26**(4), July–August, 4–8.

Oculi, O. 1979. Dependent food policy in Nigeria 1975–1979. *Rev. Afr. Polit. Econ*., no. 15/16, 63–75.

Oguntoyinbo, J. S. 1970. Irrigation and land reclamation projects in Nigeria. *Nigerian Agric. J*. **7**, 53–69.

Palmer-Jones, R. 1981. How not to learn from pilot irrigation projects: the Nigerian experience. *Water Supply Manag.* **5**, 81–105.
Proffitt, N. 1980. Africa – the grim famine of 1980. *Newswk Int.*, August 25.
Prothero, M. 1974. Some perspectives on drought in north-west Nigeria. *Afr. Aff.* **73**, 162–9.

Scott, E. P. 1979. Landuse change in the harsh lands of West Africa. *Afr. Stud. Rev.* **22**, 1–24.
Sen, A. 1981. *Poverty and famine, an essay on entitlement and deprivation.* ILO Monograph. Reviewed in *Development Forum* **9**, 2. International Labor Organization.
Shandan, B. and M. Watts 1979. Capitalism and hunger in northern Nigeria. *Rev. Afr. Polit. Econ.* no. 15/16 (May–December), 53–62.
Shear, D. and R. Stacy 1975. The Sahel – an approach to the future. *War on Hunger* **9**, 7–14.
Sheets, H. and R. Morris 1974. Disaster in the desert. *Issue* **IV**(1), 24–43.
Simon, J. L. 1981. World food supplies. *Atlantic Monthly* **248**, 72–6.
deSouza, A. R. and J. B. Foust 1979. *World space economy.* Columbus, Ohio: Merrill.
de St Croix, F. 1945. *The Fulani of northern Nigeria.* Lagos: Gregg International (reprinted in 1972).
Stenning, D. 1959. *Savanna nomads.* Oxford: Oxford University Press.
Swift, J. 1973. Disaster and a Sahelian nomad economy. In *Drought in Africa*, D. Dalby and R. Church (eds). Report of the 1973 Symposium, Center for African Studies, University of London.

Taylor, P. 1981. *Mali.* Reported on MacNeil/Lehrer Report, Air date April 17. Transcript produced by Journal Graphics, New York.

Wade, N. 1974. Sahelian drought: no victory for Western aid. *Science* **185**, 234–7.
Wallace, T. 1980. Agricultural projects and land in northern Nigeria. *Rev. Afr. Polit. Econ.*, no. 17, 59–70.
Watts, M. 1977. Conference on the aftermath of drought in Nigeria. *Savanna* **6**, 81–3.
Watts, M. 1979. Capitalism and hunger in northern Nigeria. *Rev. Afr. Polit. Econ.*, no. 15/16, 53–62.
Winstanley, D. 1978. The drought that won't go away. *New Scient.* **80**, 166–7.
World Bank 1980. *Population, per capita product, and growth rates.* World Bank Atlas. Washington, DC: World Bank.
World Bank 1981. *Accelerated development in sub-Saharan Africa: an agenda for action.* World Bank Report. Washington, DC: World Bank.
Worthington, L. 1981. Holding the Sahara at bay. *Agenda, Agency for International Development* **4**, 16–17.

2 *Power, prosperity, and social inequality in Songhay (1464–1591)*

LANSINÉ KABA

Examples of centralized political systems designed to create more effective techniques for controlling large populations and promoting trade and agriculture in West Africa from the 4th century AD included ancient Ghana in the Sahel, Mali in the upper Niger region, and Songhay in the middle Niger valley (Mauny 1961, Bovill 1968, Levtzion 1973). A large and complex empire, Songhay functioned with little strain between 1464 and 1583 despite its location in the Sahel and despite a dynastic change in 1493. This chapter will present an aspect of development in Songhay and will discuss social structures under Sonni 'Alī and the Askiya dynasty from 1464 to 1591. Through examples drawn from the Timbuktu chronicles of the 16th century and especially those pertaining to large estates, it will explain the origins and forms of the inequality that operated at the economic, political, and social levels (Houdas 1964a). It also shows how Songhay promoted a thriving agriculture despite harsh conditions and natural calamities. The chapter concludes with an effort to determine the dominant mode of production. The thesis of the chapter is twofold. First, it indicates that property and dependence relations may be viewed as historical processes although they do not necessarily turn into violent class conflict. Second, it reinforces the view that an effective and efficient political order can lead to economic growth and prosperity.

Songhay's rise represents a high point of Western Sudanese civilization and an example of elaborate social differentiation. Moreover, by comparison with ancient Ghana and Mali, its history is relatively well documented. The *Ta'rikh al-Fāttāsh* and the *Ta'rikh as-Sūdān*, chronicles written by Songhay historians, provide most of the data on which this study is based. Despite the anomalies of the *Ta'rikh al-Fāttāsh*, it contains pertinent economic information and remains the primary source of Songhay's social history.[1] Supplemented by Abderrahmān as-Sadi's *Ta'rikh as-Sūdān*, it pinpoints strong trends toward political centralization, social differentiation, and the information of large domains. In other words, despite the lack of sufficient cadastral and fiscal information, these two works help us to understand how the process of combining free commoners and conquered peoples into a large grouping of dependent farmers and service-performing caste-like communities involved in urban and rural works reached a high level, and how the ruling groups increased their unity and reinforced their hold over the whole society. These two processes of political unification and social inequality were associated with a growth of trade and the spread of Islamic values in the cities. To a greater extent than in either Ghana or Mali, Islam, as will be shown, provided the force and the ideology that validated the subjugation of the conquered. Songhay's hierarchical patterns of dominance would indicate how a social group could hold power and control economic surplus.

Military foundations of development and inequality

The rise of Songhay as the dominant kingdom in the Western Sudan in the mid-15th century was achieved at the expense of the Mali and Mossi kingdoms, Hausa city-states, and the Tuareg, Fulah, and Bambara communities by a succession of able leaders of the Sonni and Askiya dynasties. Sonni 'Alī (1464–92), whom the oral tradition calls 'Alī the Great (*Alī-Ber*), was the initiator of Songhay's grandeur. 'Always victorious never defeated', he expanded the petty kingdoms of Gao into an empire covering a large portion of West Africa and southern Sahara (Boulnois 1954; Houdas 1964a, p. 82). His religious syncretism and his opposition to the intrigues of Timbuktu's dominant Muslim faction resulted in a civil war after his death in 1493, and subsequently in the deposition of his son Sonni Baru by the Muslim forces headed by Muḥammad Toure (Hunwick 1964, Levtzion 1977, Kaba 1978). Sonni 'Alī's successes, in part, depended on his intrinsic qualities and his creation of a strong army. Under him, military achievements became part of the dominant social values and criteria for upward mobility. In 1464, the army was probably small. Conscripts usually gathered for specific campaigns and then disbanded. Sonni 'Alī introduced a quasi-permanent army. All adult males physically able to bear arms were drafted; specialized units including police forces, royal guards, grooms, and other paramilitary caste-like groups became part of the Court's own troops. The army became thoroughly structured and more effective. The strength and the preparedness of the armed forces were viewed as essential to the security of the state. To a degree, this strength relied upon the Songhay people's established traditions of chivalry: they were 'very experienced in the art of warfare and strategy, brave, intrepid, audacious, and gifted in the tricks of war' (Houdas 1964a, p. 146).

As militarization became a major trait of Songhay, competent soldiers could build brilliant careers and improve their social position. Within this context, militarization intensified the consciousness of social hierarchy and contributed to further status differentiation. Military leadership was predicated on both achievement and ascriptiveness in the sense that commoners as well as aristocrats, and even some slaves, could obtain appointments to the prestigious offices of ministers, governors, or district commanders (*farma*) generally given to officers.[2] Slaves were drafted mostly into the infantry, and their appointment to high political and military command may be explained in terms of the growth of the state and the need of the Crown for competent and loyal servants. As Olivier de Sardan has shown, these slaves could hold prestigious offices and command respect from all regardless of their backgrounds because they represented the royal power (de Sardan 1975). Militarization increased further after the revolution which brought about the Askiya dynasty in 1493. Unlike Sonni 'Alī who drafted all able-bodied adult males, the Askiya Muḥammad established a professional army consisting of a large cavalry of noblemen, an infantry of commoners and slaves, and paramilitary units directly attached to the Court (Houdas 1964a, p. 118). He also initiated the rule by which the king inherited the properties of his foot soldiers (Houdas 1964a, p. 211). His successors continued to practice these policies. For example, according to Kāti, the royal guard doubled during the reign of the Askiya Muḥammad Bunkan (1531–37) (Houdas 1964a, p. 159). The army continued to increase during the period 1493–1591 because all the Askiyas waged long wars.

Military and political expansion resulted in a phenomenal increase of the slave population whose labor was the foundation for most of the economy. The Askiya Dāwūd (1549–83) implicitly acknowledged the relation between war and large-scale slavery when he affirmed that he had been very fortunate to inherit 500 slaves from a defunct servile overseer without waging war (Houdas 1964a, pp. 193–4). The growth of coercive institutions could reinforce social differentiation. In Songhay specifically, militarization sharpened the slave–nonslave dichotomy by reducing many slaves almost exclusively to work in households and in the productive sectors for the Crown and the notables. It led to the rise of a plantation economy and of an elite whose economic and social status increasingly differed from that of other members of the society while depending on slave labor. The most significant difference between Sonni 'Alī's regime and that of the Askiyas was the forced imposition of some Islamic principles upon the state, and the recognition by the Askiyas of Islam as a system most useful to their state. This change would affect the scope of inequality.

The role of Islam

With the dynastic change of 1493, there emerged perhaps one of the most pervasive and dogmatic ideologies in the history of the Western Sūdān before the era of Islamic revival (*jihād*) in the 19th century. It affected most aspects of social organization because of its theological and militant origin.

Indeed, under the Askiyas, Islam provided support to military expansion and condoned enslavement. Unlike the Sonni regime which sought to control Muslim influence, and to a greater degree than in Mali whose kings (*mansa*) maintained an equilibrium between Islam and folk religion, the Askiya state was aggressive and intransigent in its religious policies (al-Omari 1927, Hunwick 1964, Niane 1975, Levtzion 1977, Kaba 1978). The regime viewed conversion as a noble cause. Becoming an ideology of domination, Islam institutionalized the methods by which the elite controlled the society, and perpetuated the social relations through which this dominance was carried out. With the establishment of the Askiya dynasty, the major problem for dynastic leaders was both one of consolidation of power and one of legitimacy. Although Songhay might not be a Muslim theocracy *stricto sensu* (Zikria 1958, Gibb 1962, Rosenthall 1968, Gardet 1969, Watt 1971, Khaldūn 1974),[3] Islam enabled the new regime to solve these problems and to deal with the related issue of freedom and slavery.

First, the intelligentsia (*'ulamā'*) contrasted Sonni 'Alī's attitudes with those of the Askiya Muḥammad. As the two *Ta'rikh* and the shaykh al-Maghīlī's sentences show, they disparaged 'Alī's policies and image, and ennobled those of the Askiya.[4] Thus before 1493, most of the *'ulamā'* obligated the Muslims to overthrow pagan or misbelieving rulers and to seize power in Islam's name. In Pathé Diagne's words, the rise of the Askiya dynasty marked a special problem in terms of legitimacy and sovereignty by stressing an explicit and violent claim of power in the name of the Muslim community and by delimiting the foundations of sovereignty (Diagne 1967, p. 166). As a rule, it was understood that the *'ulamā'*, because of their capacity of rational judgment (*ijtihād*), were the representatives of the people accredited to the 'commander of the faithful' (*amir al-mumin*; or *calife*), the title which the Askiya Muḥammad received after his pilgrimage in 1497. In sum, the literati questioned the legitimacy of the old regime to valorize the Askiya rule.

Because of several contradictions, the role of Islam in social relations, although far from insignificant, may not be as apparent as in constitutional matters. Islam reaffirmed slaves' legal status as bondsmen, while advising that they be treated with lenience and be manumitted after a reasonable period of service (Gardet 1969, Fisher & Fisher 1971, Lewis 1971). On the one hand, the issue was how to safeguard the principle of ownership and the right of the owners, and on the other, how to recognise the humanness of all including the servile groups. Islam elevated slaves' status beyond that of mere properties by considering them as full-fledged human beings but devoid of political rights and submissive to the will of those 'not so fortunate individuals entrusted by God with their welfare'.[5] Slaves thus constituted a group having strict obligations. The *'ulamā'* accommodated themselves to this view of slavery and Islam. Conversely, Islam became the spiritual foundation of the new social order by providing the state with a coherent ideology and by establishing greater continuity in the process of social domination while contributing to the elite's wellbeing. In other words, the Islamic doctrine enabled a minority of the population to have greater access to power. From the militant Muslims' point of view, a principal condition to hold power is to be a Muslim and to acknowledge the primacy of the Qu'ran and canon law (*sharī'a*). The head of the state must also make his decisions consistent with the dogma through the advice of the *'ulamā*.' Furthermore, Muslim authority may be imposed upon non-Muslims by force if necessary, in order to convert them and integrate them into 'the city of God' (Gardet 1969).

As a principle, non-Muslims lacked the 'constitutional' guarantees given to Muslim 'citizens', although such monotheistic groups as the Jews living near Timbuktu and in the portion of the Sahara under Songhay rule enjoyed some freedom and protection as religious minorities (*dhimma*). On the whole, the Askiya dynasty's views conformed to this interpretation of equality and tolerance. Indeed, the draconian anti-Jewish measures edicted by the well known shaykh Abd-al Karim al-Maghīlī had little effect on the Askiya Muḥammad and his advisers (MBaye 1972, Batran 1973). However, there was a consensus about the status of the polytheists. In theory, these had no political rights in the Askiya state. Pagans belonged to the 'realm of war' (*dar al-harb*). As such, only relations of conflict *de jure* existed between them and Muslims, and hence they could be enslaved. This explains why some *'ulamā'* gave a religious sanction to some of the Askiya's expeditions, which were hastily called *jihād*, although they were not markedly different from ordinary wars and slave-raidings. Examples would include the campaign that the Askiya Muḥammad made in 1498 against Naasira, a Mossi king, with the shaykh Mursālih Jawara to 'make it a true holy effort' (Houdas 1964a, pp. 134–5, & 1964b, pp. 121–3). Naasira having refused to be converted to Islam, the Askiya attacked, and seized countless booty and captives, some of whom were settled as royal slaves along the Niger River. Therefore, it would seem that the *'ulamā'* condoned slavery as Olivier de Sardan has argued (de Sardan 1975).[6] With economic prosperity, the contradictions between religious ideals and material gains further increased as is evident in the case of the rich overseers who, although converted, were not freed (Houdas 1964a, pp. 179–99).[7] Scholars justified slavery as a system of social control and a form of ownership consistent with the traditions of Islam and Songhay.

To underscore this idea of continuity further, suffice it to recall how Islamic ideology remained strong in Songhay despite the deposition of the Askiya Muḥammad in 1528. The succession crises had little effect on the position of Muslims; and no monarch challenged this policy. To the contrary, the Muslim influence increased further, and the Islamic dogma compatible with both private ownership and the state's participation in the economy (Rodinson 1966, 1968) contributed to the rise of large estates, and hence to the division of rural populations into free village communities and servile groups including the laborers and the overseers. The development of these huge manors and their need of labor partly explained some of the ideological ambivalence of the Muslim elite in matters of social relations.

The rise of a prosperous intelligentsia

During the period 1464–1591, Muslim chroniclers usually judged a reign by its policy toward Islam and its generosity toward the *'ulamā*.' According to S. M. Cissoko, after 1493 the sovereigns remained loyal to the tradition of 'alliance between the throne and the altar' (Cissoko 1975, p. 92).[8] Not only did this interdependence benefit the two partners – the Crown and the intelligentsia – but it also deepened social divisions in general and even among the elite as evidenced by the disagreements between the *'ulamā'* and some civil administrators over tax exemption for the scholars (Tymowski 1970). The link between politics and religion was both a factor of wealth and influence for the beneficiaries and a process partly responsible for the flourishing of a brilliant urban civilization. This made Songhay 'one of the countries most favored by God with riches and abundance, peace and security, beauty and splendor in knowledge and piety, purity of customs, safety of properties, compassion toward the poor and the foreigners, and respect and fellowship for scholars and students' (Houdas 1964a, p. 313. For the rise of a prosperous elite, see Saad 1979). Exempted from observing most etiquette, the *'ulamā'* took precedence over other dignitaries at the court in Gao, and some of them also received appointment to high offices, in particular judgeships and positions of royal advisers. These honors were accompanied by a policy of granting large fellowships and donations, a policy that Sidi M. Kāti attributed to the Askiya Muḥammad.[9] The Askiya's philanthropy was both spiritual and material. In the spiritual realm, he helped many scholars to accomplish the pilgrimage to Mekka (*hājj*), a canonical obligation resulting in a highly valued title. In August 1467, the Askiya Muḥammad went to Mekka with an escort of dignitaries including seven eminent scholars whom he treated with great deference. In Arabia, he gave alms amounting to thousands of gold dinars and bought two buildings to be used as religious foundations (Houdas 1964a, pp. 25–7 & 131). Philanthropy became a dominant feature of the regime.

A few remarks on the nature and the amount of some donations may further illuminate the level of agriculture and the magnitude of the relations between the Crown and the *'ulamā*,' although the figures must be taken with reservation. In 1501, after a major victory in the region of Agades, the Askiya offered to his adviser, the shaykh Ṣāliḥ Jawara, an estate near Gao which included three groups of Crown slaves called *zanzi*, who differed from ordinary slaves in that 'they had never at any time been free people and could not be sold or manumitted but by the Askiya, or purchase manumission or marry free persons' (Hunwick 1970). To the eminent shaykh Muḥammad Tule, he gave the 'usufruct of the area which a horse rider can cover from sunrise to sunset' from a location called Harkunsa-Kaiguru near Gao, and which included three *zanzi* clans and many farms. Once in 1507, the Askiya, upon encountering the shaykh Mori Hawugaru's offspring in Timbuktu, gave 10 slaves and 100 cows to each, promised to renew this gift every year, and exonerated them from paying tax (Houdas 1964a, pp. 136–40). Noticeably, some of the most spectacular gifts recorded in the chronicles went to foreign scholars who claimed a *sharīf* origin,[10] for example, the sharīf Mulāy Ahamd al-Saqlī who received many impressive gifts (Houdas 1964a, pp. 37–8, 198–201 & 212–14) upon moving from Mekka in 1519.

The Askiya Muḥammad's successors in general, and the Askiya Dāwūd (1549–83) in particular (Houdas 1964a, ch. XI), practiced the same policy. Dāwūd engaged scribes to duplicate manuscripts for the teachers in order to improve their libraries. Each year during his reign, he sent 4000 *sunnu* of grain to the *qādī*, the head of Timbuktu's Muslim community, for the students and the needy. (The *sunnu* is a leather bag containing 150–180 lb.) He also maintained there a farm with 30 slaves to supply vegetables. As a rule, he distributed part of the spoils and tributes among scholars and religious institutions. For example, he gave out of charity the vast heritage of the slave head, the *jango* Musa Sagansaro to teachers and mosques for 'God's love'.[11] In a different context, the writer Alfa Mahmūd Kāti received 40 *sunnu* of grain, a farm with 13 slaves and many cows, and 80 gold *mitqals* to buy a rare book. (The *mitqal* equals about one-eighth of an ounce.) Finally, in *circa* 1581, to atone the accidental killing of the sharīf Muḥammad Muzāwir, son of the shaykh al-Saqlī, the Askiya fasted each day until his death in 1582, and gave a restitution which amounted to 30 times the customary rule and which included three *zanzi* villages of 200 slaves, each located along the Niger River (Houdas 1964a, pp. 212–15). Undoubtedly, the prodigality reinforced the patrician families' position in the cities, and made them an affluent group (Saad 1979). They recovered the property and the prestige that they had lost under Sonni 'Alī. Although all the Askiyas were not devout Muslims, they acknowledged the pre-eminence of Islam and secured the needs of the Muslim notables.

The Askiyas' philanthropy brought about a change in land tenure and agrarian structures (Houdas 1964a, pp. 108–13). Previously, conquered and perhaps unexploited lands belonged to the Crown. For example, the Askiya Muḥammad held the title to all 24 of the servile clans and their land that the Sonni dynasty took from the *mansa* of Mali after Songhay's liberation (Houdas 1964a, p. 106). These communities, who usually cultivated plots of about 40 m^2 per married couple for the *mansa* and provided for their own food, were under the Sonni to till parcels of about 200 m^2 in gangs of 100 men placed under the direction of rural managers (*fanfa*, pl. *fanāfi*). After the overthrow of Sonni Barou in 1493, this system became relatively more liberal (Houdas 1964a, p. 109).[12] Prestations were set at 10, 20, or 30 bags of flour maximum per family according to their means. This reform meant the end of the old division of Crown land into lots for the Askiya and lots for the laborers. It also involved greater land concentration for the Askiya Muḥammad as well as greater possibilities for the families to produce for themselves by controlling their surplus rather than the land. The reform probably resulted in a reduction of the total dues in kind paid by individual slave families although the growth of the servile population could compensate this reduction. Tymowski's thought that it was not a radical change may be correct (Tymowski 1970). Indeed, it reduced oppression, and thus was beneficial to the producers, while it perpetuated the old system of social relations as evidenced by the strengthening of the Askiya's prerogatives: he held the land titles and had rights over the slaves, as he could take and sell their children to buy horses or could use them as stablemen (Houdas 1964a, p. 109).

The effects of this reform, and especially the decline in dues, could not escape the attention of competent administrators, in particular of the governor of the rich farming province of Kurmina, the future Askiya Dāwūd. This Askiya imposed a more rigid system which brought about more dues in kind (Houdas 1964a, pp. 178–88), and hence a larger labor force and more production. These two processes significantly stimulated the economy. Despite the differences between the Askiya Muḥammad's policies and those of his son, it would seem that a substantial amount of resources (land and/or labor) were transferred from the Crown to the religious elite and to institutions. The literati consolidated their economic position, although some donations excluded titles to the land. The beneficiaries could involve their slaves in various economic activities, including trade, animal husbandry, and the emerging rice farming in floodplains. As a result, not only did the scholars strengthen their hold on the ideology of the state, but also they associated themselves with other dominant strata through their participation in urban politics and trade.

In general, despite the years of drought and epidemics, the Askiya dynasty witnessed a relatively long period of peace and prosperity. Security, by stimulating production and exchange, contributed to the growth of individual wealth. Long distance trade within the Sudan and with North Africa flourished so much that the Askiya Dāwūd had to build a storehouse for his gold coins (Houdas 1964a, p. 177), and trade became so important that the eminent shaykh Sidi Yaḥyā and other scholars actively practiced it (Houdas 1964a, p. 82). To a degree, this process symbolized the emergence of the *'ulamā'*-businessman, a process consistent with the role that business interests played in Timbuktu politics before 1493. The link between worldly and spiritual activities was particularly evident in the tailoring craft, a profession free of caste stigma and hence open to free men and scholars.[13] At the time of the Askiya Dāwūd, 26 tailoring houses functioned in Timbuktu, each with an average of 50 apprentices who were also Qur'anic students (Houdas 1964a, p. 315). Master tailors often numbered among the well educated people. Such activities and the exchange of local and regional surpluses were facilitated by the military expansion and the subsequent control by Songhay of a large market having its axes centered approximately around Taghaza to the north, Walata to the northwest, Jenne to the southwest, Katsina/Kano to the southeast, and Dendi to the south – Timbuktu and Jenne were the main metropolises.[14] To a large extent, it was a big business using efficient means of land and river transport. Caravans, using both animal and human power, connected different localities regularly, quickly, and in great numbers. Significantly, river transportation, a consequence of the naval policy, became vital as evidenced by the involvement of the Askiyas and many notables in it.[15] Most of Jenne wholesalers also owned large barges for carrying passengers and bulky products. Hence transportation became dependable and affordable. Gao, with its twin ports of Guyima and Gadayi; Timbuktu, with its port at Kabara; Jenne, with its dockyards and Ras al-Mā on Lake Faguibine, all became major trading centers with a prosperous elite. The trade also involved multilateral relations across the Sūdān and the Maghrib and was even in contact with the emerging European mercantilist economy through the Arabs. Slaves represented a significant part of the export commodities to North Africa. However, as a state necessity along with gold, slave trade operated under some control, as implied in the Askiya Dāwūd's refusal to sell to an Arab the 500 slaves inherited from a defunct overseer.[16] It would seem that, though practiced, it was not dominant in the 16th century given Morocco's need of gold and the Songhay policy of forced settlement. Yet there was a demand for slaves in Morocco, probably for sugar cane production.[17] Thus warfare in Songhay might not withdraw men from production, although it could at times be complementary to export trade. By the mid-16th century, the Moroccan Sa'did dynasty had placed the Sūdān at the center of their overall policy because of the pivotal role of the trans-Saharan trade. Above all, the sultans wanted to find more gold to finance their trade with Europe (Braudel 1949, Kaba 1981). Despite the excellent conjuncture coinciding with the Askiya Dāwūd's reign, the trans-Saharan gold traffic could not satisfy Morocco's needs because of the establishment of European owned and managed comptoirs on the coast of West Africa.

Within the Sūdān, these relations resulted in a greater interdependence between regions producing complementary good: that is, farm products and fish from the Bend of the Niger River; gold from the south; horses, books, weapons, textiles, salt, and copper from the Maghrib and the Sahara.

These activities further increased prosperity and enabled the privileged to have access to more imported luxury goods. Islam provided the unifying principle and the moral code necessary for the success of this large-scale commercial integration. Muslim prayers and universalism promoted the sense of community, thereby helping to create the conditions necessary for trust and credit. Despite the significance of trade, however, the number of merchants in Songhay was probably not very large. Therefore its social impact was limited because of the *de facto* monopoly of Arabo-Berber businessmen over the lucrative northern section of the trans-Saharan trade and because of the very nature and size of the clientele. To quote A. G. Hopkins, 'long-distance trade tended to cater to the needs of relatively high income groups able to pay prices which took account of the scarcity-value of items that were unavailable locally' (Hopkins 1974, p. 58). Because this clientele was mostly interested in prestige goods and conspicuous wealth, the trade was rather an exclusive business. Therefore, to understand inequality in Songhay and the nature of the society better requires a look at rural activities.

Agriculture and the nature of society

From Sonni 'Alī's time on, agriculture may be viewed as the dominant sector of the economy with major rural projects including hydraulic works and land colonization in Masina, Gurma, and Timbuktu regions. Indeed, Gaston Mourgue has found numerous remains of a relatively large sedentary population spread throughout the region. Hydraulic works built near villages, and designed to retain rain waters or to tap underground pools, would testify to a definite know-how and to an extensive agricultural need. These works included trenches for irrigation, dikes, wells, and reservoirs.[18] Jewish communities living in Tendirma also had deep wells to produce premium quality vegetables with profit (Houdas 1964a, pp. 119–21). The scope of the rural projects made the state a dominant force in production. This economic conjuncture was maintained despite the drought and the plague which brought havoc in 1549 and 1582.

Thus the Askiya had almost everywhere large estates using forced labor and yielding vast amounts of crops, including rice (*mogho*), millet (*subu*), barley (*farba subu*), and even wheat (*alkama*). Millet and wheat were the main and complementary staples, respectively, whereas rice remained a luxury. Units consisted of labor gangs working under appointed overseers and could number from 20 to 100 families (Houdas 1964a, pp. 179 & 211). Michal Tymowski's suggestion that 'family' here means a married couple may be accurate, although polygamy need not be viewed as incompatible with this system (Tymowski 1970, p. 1649) since the Askiyas were interested in large slave families because of their ownership rights over the children.

Two main agricultural systems probably existed in 16th-century Songhay: that of the free village communities and that of the large estates. Despite the chroniclers' silence on the structures of the free rural communities, all seems to indicate that this system was perhaps based on communal ownership and sharing. As for the second mode, which received much attention in the chronicles, it included Crown and private domains based on land colonization and involved in rice farming. A rule of production was that the Askiya each year would supply the seeds and the hides needed to make the bags. He would also send to the head *fanfa* 1000 kola nuts, a big bar of rock salt, a very wide garment without sleeves or collar known as a *boubou*, and a large wraparound skirt for the *fanfa*'s senior wife (Houdas 1964a, p. 182). Each estate would deliver a fixed amount of the crop regardless of the total annual harvest. These dues were carried in barges to Gao, Tendirma in Kurmina, Kabara, Koukiya, Bamba or other strategic centers having royal granaries. Obviously, the yield as well as the labor conditions varied according to the size of the estate. Some laborers probably lived in poverty whereas their overseers (*fanāfi*) conspicuously exhibited their wealth as exemplified by Missakulallah (see Houdas 1964a, pp. 180–1 & 189–91, for wealth of Musa Sagansaro). One may argue that slavery reduced the demand for some products, and hence reinforced the trend toward self-reliance. Yet it tended to generate an inequality of income.

The accumulation of wealth by the overseers would imply that the yields were relatively high and that the management techniques were rational enough to safeguard the interests of both the Crown and the *fanāfi*. This agriculture, to succeed, required a large concentration of slaves and relied on manipulation, coercion, and the threat of punishment. Oppression could be viewed as a trait of large estates. As the Askiya Dāwūd summoned Missakulallah to Gao to explain his strange behavior of giving out of charity all the yield of his manor, so the highest ranking intendants had rights only insofar as they performed their customary duties well. They could be arrested, chained, and imprisoned for insubordination or for the failure to live up to their obligations under the oral contract (Houdas 1964a, p. 182). To a degree, therefore, violence or the threat of it became almost an instrument of conquest and control, a privilege of the owners, and a symbol of rank and status. Hence, the Crown and the elite could secure extensive labor and good from the estates, despite recurring natural calamities. This practice, as has been noticed earlier, was a decisive factor in the rise of the brilliant, cosmopolitan 16th-century civilization and a sign of some class division and consciousness (Houdas 1964a, pp. 312–16).[19]

The nature of society Songhay consisted of various production systems including elements of communalism, slavery, and a trade-based economy. The role of slaves in production, conveyance, and services, the high involvement of the state in the economy, and the transfer of land and people from the Crown to some members of the elite, have all led to different speculations about the nature of the society. For example, Tymowski has viewed the transfer of property as a trend toward feudalism and Majhemout Diop has spoken of vassal links; D. Tamsir Niane who has viewed Mali as a centralized feudal state would also speak of Songhay's feudalism; and S.-M. Cissoko has had much difficulty in accepting that Songhay was a slave-based society (Tymowski 1970, p. 1653; Diop 1971; Cissoko 1975, pp. 169–70; Niane 1975, p. 63). Admittedly, slavery grew in Songhay between 1464 and 1591. As Cissoko wrote, 'under the Askiyas, slaves constituted the numerically dominant part of the population, and each family according to its means had a few' (Cissoko 1975, p. 56). Even slaves could own slaves, as the account of Musa Sagansaro's inheritance indicated. This means that the very concept of slavery, flexible and vague, suggests a comprehensive web of dependence relations involving the whole society from the bottom to the Crown. Everyone except the Askiya was 'the slave' of someone else; and they were all of them the servants of the monarch who, according to the Islamic hierarchy, was the chief servant of God. Slavery, as implied in the cases of Sagansaro and other overseers owning their own slaves, denotes a form of labor control and ownership that denied the freedom of those at the bottom of the bondage system. The right of the overseers to own slaves indicated the flexibility of the system and the landowners' trust rather than the freedom of the overseers. The masters' self-serving indulgence was inseparable from their right to own the progeny and the inheritance of their slaves. It also expressed a subtle process that increased the number of slaves accruable to the owners. In short, the success of the Songhay slave system relied in part on the overseers' involvement in rural activities both as officials and dependent owners with limited rights.[20]

There were two main categories of slaves: the Crown slaves, including the *zanzi* and the captives, and those privately owned. (The word 'slave' refers to any person devoid of civil rights, because of captivity in war, lawful purchase, or birth, belonging to, and working for, the state or religious institution or another individual.) Slavery varied according to the owner's social position and the responsibilities assigned to the slave as the case of Missakulallah would indicate. Given the conquests and the role of the state in production, the number of royally owned slaves probably grew to exceed the private ones. Large captured communities belonging to different ethnic groups were forced to settle and work in specific areas. To this category also belonged the 24 Bambara Sorko and Fulani servile groups whom the Sonni dynasty 'inherited' from the *mansa* or Mali, and who were to be inherited by the Askiyas in 1493 (Cissoko 1975, p. 56).[21] In general, they were associated with metal working, hunting, fishing, and riverain occupation. Divided into caste-like units with special functions, they produced crops, boats, weapons, or fodder. To these groups, the conquests of the Askiyas added an innumerable number of captives, most of whom also received hereditary professions and statuses, although different from the *zanzi*.[22] In Songhay, slavery was accompanied by a qualitative change of status which often resulted in caste-like structures and features for those at the bottom of the social hierarchy, although the society was not based on caste.[23] These structures and the inherent abasement of the subjects concerned belonged to a long process of domination and control, and denoted a closed and rational pattern of class recruitment. Through this process, the groups in bondage would receive specific hereditary obligations and characteristics, and would perpetuate themselves in these conditions in a manner that did not cut across the whole society. As a result, the Crown and the elite could secure for themselves, free of charge, the service of those whose professions were socially and economically necessary. However, at a later period and in a different context, these professions could become the monopoly and the prerogatives of the subjects' descendants practicing the craft.[24]

Furthermore, contrary to the impression given by Jean Suret-Canale (1961), slavery was severe.[25] The unity of slave families could easily be destroyed, as implied in the appeal that an old woman made to the Askiya Dāwūd, after the latter inherited herself and her 27 offspring, to sell or to offer all of them to the same person in order to maintain their unity (Houdas 1964a, p. 192). Slavery was also expanding rapidly. Thus slaves played a predominant role in rural production. For Cissoko, they formed the bulk of the population, and slavery had become the dominant mode of production in the rural areas. Yet the same author argues that the urban trade-based economy and the productive labor of the free men were more significant than slave production in Songhay (Cissoko 1975, p. 170). By contrasting the urban and rural sectors, Cissoko has implied a pertinent question about the coexistence of two modes of production, the communal free sector and that of the slaves. The lack of data on the free village communities notwithstanding, one may presume that they constituted the *old basis* of Songhay providing soldiers for the army and valuable annual tributes.

To understand the nature of the Songhay society, it is useful to consider the main traits of the state in the 16th century. For example, the growth of the state apparatus, the rise of a strong urban elite, and the growing demand for forced labor. Since land had become plentiful, the problem was how to exploit it most effectively with minimal disruption of the social structures which were based on the freedom of the rural communities. The Askiyas had to find a different source of labor to cultivate the new farms; and this could not lead but to an involuntary gang-type of labor system. Without this extensive slavery, the labor costs of large estates, transportation, and of public projects would be prohibitive. As a result, there would be only a small propertied class. To a large extent, a main part of the policy of Sonni 'Alī and the Askiyas, as has been suggested, was to prevent a labor shortage and to secure regular production and efficient control of labor. Thus they practiced an accommodationist policy with the *fanāfi*. Due mostly to their managerial skills and the relative autonomy of their positions, these overseers accumulated wealth and almost became a privileged rural group. This is why Tymowski has called them serfs (Tymowski 1970, p. 1653). But the two *Ta'rikh* insist on their slave status and the Askiyas' right to their inheritance. These restrictions limited the overseers' attempt to consolidate their positions and become a full-fledged free class. Contrary to Suret-Canale's assumptions (Suret-Canale 1961), this process would indicate that large-scale slave production dominated some sectors of the Songhay economy.

In terms of production, the contrast was growing between free village communities and the slave ones. The rural communities continued to practice the old communal forms of social relations. In terms of land tenure, Mahmūd Kāti emphasized local communities' control over their land. The advice given by the shaykh Muḥammad al-Maghīlī and the shaykh Abderrahmān al-Ṣuyuṭī to the Askiya Muḥammad reiterated the inalienability and the inviolability of such properties (Houdas 1964a, pp. 15–21; MBaye 1972; Batran 1973). At the same time, as has been shown, there was a spectacular increase in the number of Crown and private estates. Despite the large number of free villages, it would seem that their role in the economy was declining because they could not satisfy the new demands related to the growth of the state. This newer form of production involved more complex types of farming – rice production in floodplains, for example – and of relations of production and dominance. Among these forms, the Askiya's rights to the land and the people, hence slavery, were particularly significant. By 1591 when the Moroccans invaded Songhay, the new mode had not entirely superseded the old communal relations of production. Nevertheless, a class of nonproducers, the large landowners, had established a strong relationship of exploitation and inequality supported by force and religious dogma. The absence of a clear-cut mode of production, and hence the coexistence of the two modes of production would be consistent with the view that a mode of production does not itself exist in isolation (Terray 1975).

Slavery reconsidered

This idea may help us to come up with a better understanding of the nature of Songhay society and to postulate some hypotheses. To begin, despite Tymowski's commendable efforts, the data from the *Ta'rikh al-Fāttāsh* do not represent 'a vivid image of feudal relationships being formed in both the political and socio-economic realms' (Tymowski 1970, p. 1641). Neither do they support Diop's opinion that Songhay was in transition from slavery to feudalism (Diop 1971). The *fanāfi's* relations with their masters would indicate tributary links rather than feudalism. To consider Songhay a feudal state would amount to a 'tyranny of concept'.[26] A *fanfa's* legal status cannot be confounded with that of the plebeian serf, and certainly not that of the noble vassal. Moreover, there is no hint in the two *Ta'rikh* that the Askiya gave *fiefs* to his noble officers or to the *'ulamā'* in exchange for their services, and that he inherited part of their wealth. To strengthen this argument, suffice it to recall that feudal economy kept a local characteristic for a long time and that the medieval lords' power proceeded from the weakness of the Crown. Thus political fragmentation, and hence the multiplicity of the sources of political authority and administrative rule often characterized feudalism. The opposite was true of Songhay where Sonni 'Alī and the Askiyas established a relatively centralized system and kept officials under control. The economic sector involving slave labor in latifundia, mining, and long-distance trade was growing at such a fast rate that one might speak of the emergence of a slave mode of production. This trend was quite compatible with some forms of communalism.

In sum, from Sonni 'Alī's reign in 1464 until the Moroccan invasion in 1591, Songhay experienced a gradual structural change which increased production and heightened the patterns of dominance that could lead to large-scale slavery characterized by the ownership of the laborers and the control of the land by the Crown and the notables. This slave-orientated system implied that all laborers including the caste-like groups were considered as 'property', or more accurately, human resources employed in their own reproduction as hereditary units and in the production of goods. In conclusion, social inequality and the trend toward a slave-based economy in 16th-century Songhay resulted from both internal processes and economic relations with the outside world. These included the expansion and consolidation of the state, the establishment of a legitimating ideology and the uneven trans-Saharan trade. Slavery was linked to the efforts of the state and the elite to participate in this exchange with North Africa. However, the Moroccan invasion of Songhay in 1591 brought about major upheavals that ultimately destroyed the system (Kaba 1981).

Notes

1 The text of *Ta'rikh al-Fāttāsh* consists of three distinct parts written by at least three different members of the Kāti clan. The first was Sidi Kāti whose name appears in the preface and who was an adviser to the Askiya Muḥammad (see Houdas 1964a, pp. 24, 26, & 135). He began writing the chronicle in 1519 (see Houdas 1964a, p. 27). According to Houdas, he was born in 1468 (1460 according to Félix Dubois 1897) and he died in 1553; his original text is lost. According to Hunwick (1969), the second author was Mahmūd ben al-Muttawakil Kāti, a nephew of Sidi Mahmūd (Hunwick 1969); he is often mentioned in the text as Alfa M. Kāti. Adviser to the Askiya Dāwūd, he participated in the

battle of Tondibi and died in 1953. As for the third apparent author, his surname is not given in the work; he was a grandson of Alfa M. Kāti through his mother, and his father was El-Mukhtar Gumbele (see Houdas 1964a, p. 135). This third writer took extensive quotes from the version written by his grandfather Kāti (see Houdas 1964a, p. 92); he witnessed the first 8 years of the Moroccan rule. It is this third text with which the shaykh Aḥmadu of Massina tampered in the 19th century to legitimize his rule and ennoble his own image (see Houdas 1964a, pp. 17–19, for example).

2 For example, Muḥammad Toure (the future Askiya), who was reportedly a member of the royal family through his mother, served as governor in Hombori (*Tondi-farma*). The highly respected office of the minister of the fleet (*hi-koy*) was generally held by a commoner of servile background. Under the Askiya Muḥammad, many officers, in particular 'Alī Folon (a most influential adviser and once governor of Jenne), the chief of protocol and chief of the tax collectors at the port of Timbuktu, were of slave extraction.

3 Indeed, it would seem that Islamic values, although dominant in the cities under the Askiyas, did not supersede animism in the rural areas (see Boulnois 1954 and Hama 1954).

4 See Houdas, for example: 'The Askiya Muḥammad eliminated all the reprehensible innovations, inequities and bloody cruelties of the Sonni; he restored religion on the most solid bases by instituting a *qadi* and an *imam* in each town of the empire.' The religious commitment and the class nature of the Timbuktu scholarship are explicit in this quote (Houdas 1964a, p. 115). This view is consistent with the shaykh al-Maghīlī's sentences (MBaye 1972, Batran 1973).

5 This is an expression commonly heard among Qur'anic teachers in West Africa and reportedly going back to remote times. It implies the responsibility of the masters here and the hereafter, and hence their salvation or damnation.

6 Within this context, Mamadou Dia has asked a very pertinent question: 'Si l'Islam est une religion de la personne, de l'égalité et de la justice, comment expliquer l'esclavage . . .? De telles institutions ne sont-elles pas la négation flagrante de tout personnalisme qui se donne un contenu?' (Dia 1977, p. 79).

7 The examples would include Missakulallah and Musa Sangansaro whose conditions will be discussed later (Houdas 1964a, pp. 179–99).

8 As examples of how the literati judged the monarchs, Muḥammad-Bunkan (1531–7) was praised for his respect for the scholars' advice despite his inclination to ostentation and musical entertainments; the Askiya Ismael (1537–9) received unequivocal blessings for freeing the aged Askiya Muḥammad I; the piety and philanthropy of the Askiya Ishaq I (1539–49) impressed the historian Sidi Mahmūd Kāti. But no other Songhay sovereign after the Askiya Muḥammad embodied better than the Askiya Dāwūd (1549–83) the virtues that the Muslim literati admired in statesmen. Alfa Mahmūd ben al-Mutawakil Kāti described Dāwūd in the most laudatory terms (see Houdas 1964a, pp. 176 & 207, and for more details, pp. 37–8, 198–201, & 212–14).

9 M. Kāti wrote: 'He was extremely fond of the "*ulamā*," the holy men and the students. He usually gave many alms and accomplished many supererogatory acts of devotion in addition to the prescribed ones. Full of attention to the "*ulamā*," he generously distributed slaves and riches to them to ensure Muslims' interests and assist them in their religious practice and their submission to God' (Houdas 1964a, pp. 114–15).

10 A *sharīf* is a person claiming descendance from the prophet Muḥammad.

11 This heritage included 500 slaves, 1500 bags of grain, 7 herds of cattle, 30 flocks of sheep, 15 horses including 7 thoroughbred chargers, and countless weapons and other items (Houdas 1964a, p. 191).

12 Indeed, these servile families could produce and keep more crops for themselves.

13 Free men and even notables could practice tailoring because textiles represented a significant portion of export commodities from abroad. Their business was an activity much dependent upon the participation of the elite. Furthermore, the tailoring craft

usually accompanied retail trade and tended to provide the travelers with additional funds while they were awaiting sales opportunities.

14 The conquest of Timbuktu in 1468, Jenne in 1476, and Walata in 1483 contributed to this broad integration. To understand the economic importance of this region, note that the Mossi king's attempts to extend his rule over Walata and the Niger Bend led to a series of brilliant counter-offensives by Sonni 'Alī between 1477 and 1483; these victories resulted in Songhay's hegemony (see Houdas 1964a, pp. 92–3 & 115). Along with Arawane, Walata was a major commercial entrepôt and the abode of famous '*ulamā*' and notables. Its strategic position contributed to the flow of goods, ideas and people between the Maghrib and the Sūdān, and hence to Timbuktu's own prosperity (l'Africain 1956). As for Timbuktu, it was successively ruled by the governor (*farin*) appointed by the *mansa* of Mali probably since the rule of Sakura (1285–1300); the *Timbuktu-koy* appointed by the Berber Magcharen (1433–68); and finally by the Songhay (see Houdas 1964a, pp. 315–16, for its prosperity, and Cissoko 1975; see Houdas 1964a, pp. 22–8, for pertinent information on Jenne).

15 For example, the *Guyima-koy*, a head of the sapper corps, maintained hundreds of big barges (*kanta*) for the Askiya at the twin ports of Gao (Houdas 1964a, p. 270; see also Tymowski 1967).

16 For the sugar production in 16th-century Morocco, see Berbrugger (1862) and Berthier (1966).

17 This Arab businessman Abdalwāsi al-Masratī lived in Gao; he offered 5000 *mitqal* to the Askiya for the 500 slaves (Houdas 1964a, p. 193).

18 Military conquest brought about the enslavement of masons and other craftsmen who worked on these projects: for example, 500 masons were captured in one expedition (see Houdas 1964a, pp. 106–12 & 118–24). As for hydraulic works, see Rouch (1953, p. 182).

19 These pages are an eloquent portrayal of the prosperity and humanistic civilization achieved before the Moroccan invasion. Class consciousness is implicit in the way the author as a representative of the '*ulamā*' refers to 'liberty', and earlier how the Askiya Dāwūd debases Missakulallah as a 'slave of humble and miserable condition' (Houdas 1964a, p. 182).

20 A careful reading of the anecdote about Missakulallah would corroborate this view, as he insisted on his loyalty and the need of trust between masters and slaves (Houdas 1964a, p. 183).

21 Levtzion has questioned the authenticity of these passages of the *Ta'rikh* (see Levtzion 1971, pp. 571–93 & note 68 for more discussion).

22 The innate status of the *zanzi* has been servitude for time immemorial.

23 Caste system is holistic. One caste cannot exist in an otherwise casteless society. In other words, castes are very interdependent social phenomena (Dumont 1960).

24 This evolution may explain Levtzion's doubt (Levtzion 1971), since caste-like groups were free in the 1820s when the shaykh Aḥmadu of Massina tampered with the *Ta'rikh al-Fāttāsh*. As A. Hampaté Ba and Daget have shown, this shaykh was very interested in maintaining the differences between free and servile persons, and claiming ownership of the slaves for the state. J. O. Hunwick's observation that the passages about the *zanzi* might not be an addition is pertinent. Indeed, the author M. Kāti stated in his preface that his work aims at 'differentiating clans of noble free condition from the servile ones' (Houdas 1964a, p. 11). Hunwick has also suggested that these offending passages might have been excised from the chronicle in the post-1591 upheavals which drastically altered the patterns of social relations and elevated the lower class above the nobility (see Houdas 1964a, p. 308). The shaykh Aḥmadu might have restored the passages to legitimize further his rights over the servile groups. Olivier de Sardan has viewed an anachronistic tendency in Levtzion's reservation about the *zanzi* (see Hunwick 1969, p. 103, and de Sardan 1975, pp. 107–8).

25 Jean Suret-Canale has argued that 'Africa never experienced the absolute forms of slavery

(which were) proper to classical Mediterranean Antiquity or to modern colonial America' (Suret-Canale 1961, p. 103).

26 The 'tyranny of concept' has made it difficult to agree on a definition of feudalism. The question is whether the hierarchy of ownership and social relations concerning the *fief*, or the manorial system binding the serfs to the estate, or simply the existence of latifundia constituted the main traits of feudalism. Between the era of slavery marked by the ownership of a person by another, and the modern capitalist era of private ownership and free labor, most of Europe went through a feudal period during the Middle Ages. As a social and economic system based on the ownership of the land and characterized by dependence links of vassalage, links which themselves were established on the promise of serving an individual as a lord (*recommendation*), feudalism generally involved several processes. These included the act of homage, the attribution of an estate (*fief*) to the vassals by the lord, and the lord's right to a portion of the crops on the fief and part of the vassal's inheritance. All groups, from the king to the nobility and from the nobles to the serfs who were bound to the estate and responsible for production, were affected by these social and economic ties. The manorial system stemmed from this organization. Thus feudalism involved social and military obligations and clearly defined patterns of hierarchical relations. For an overview of the difficulty in defining feudalism, see the prefaces of Marc Bloch's *La société féodale* and Robert Boutruche's *Féodalité et seigneurie* (Bloch 1959, Boutruche 1959); Goody (1963); and Brown (1974). Jack Goody has convincingly rejected the idea of an African feudalism on mere structural similarity because the technological aspects of feudal society lacked in Africa.

References

l'Africain, J. L. 1956. *Description de l'Afrique* (trans. A. Epaulard), vol. 2, 464–5. Paris: Adrien-Maisonneuve.

Batran, A. a.-A. 1973. A contribution to the biography of Shaikh Muḥammad ib Abd al-Karim al-Maghīlī. *J. Afr. Hist.* **14**, 381–94.

Berbrugger, A. 1862. La canne à sucre et les chérifs du Maroc au 16e siècle. *Rev. Africaine* **6**, xxxii.

Berthier, P. 1966. *Les anciennes sucreries du Maroc et leurs réseaux hydrauliques*. Rabat: Imprimerie Marocane.

Bloch, M. 1959. *La société féodale*. Paris: Albin Michel.

Boutruche, R. 1959. *Féodalité et seigneurie*. Paris: Aubier.

Bovill, E. W. 1968. *The golden trade of the Moors*. London: Oxford University Press.

Braudel, F. 1949. *La Méditerranie et le monde Méditerranéen à l'époque de Philippe II*, 368. Paris: Presses Universitaire de France.

Brown, A. R. 1974. The tyranny of a construct: feudalism and historians of medieval Europe. *Am. Hist. Rev.* **70**(4), 1063–88.

Brun, J. 1914. Notes sur le Ta'rikh el-Fettach. *Anthropos*, 590–6.

Cissoko, S. M. 1975. *Tombouctou et l'empire Songhay*. Dakar: Nouvelles Editions Africaine.

Dia, M. 1977. *Essais sur l'Islam*. Vol. 1: *Islam et humanisme*. Dakar: Nouvelles Editions Africaine.

Diagne, P. 1967. *Pouvoir politique traditionnel en Afrique occidentale*. Paris: Préscence Africaine.

Diop, M. 1971. *Histoire des classes sociales dans l'Afrique de l'ouest: le Mali*, 21–3 & 77–8. Paris: Préscence Africaine.

Dubois, F. 1897. *Tombouctou, la mystérieuse*. Paris: Le Figars.

Dumont, L. 1967. *Homo hierarchicus*. Paris: Gallimard.

Engels, F. 1972. *The origin of the family, private property and state.* New York: Pathfinder Press.

Fisher, A. G. B. and H. J. Fisher 1971. *Slavery and Muslim society in Africa.* New York: Anchor.

Gardet, L. 1969. *La cité musulmane: vie sociale et politique.* Paris: Vin.

Gibb, H. A. R. 1962. *Studies on the civilization of Islam.* Boston: Beacon Press.

Goody, J. 1963. Feudalism in Africa? *J. Afr. Hist.* **4**, 1–18.

Hama, B. 1954. *L'empire de Gao: histoire, coutumes et magie.* Paris: Préscence Africaine.

Hopkins, A. G. 1974. *An economic history of West Africa.* New York: Columbia University Press.

Houdas, O. 1964a. *Ta'rikh al-Fāttāsh* (French trans. and Arabic text). Paris: Adrien-Maisonneuve (original edn 1913–14).

Houdas, O. 1964b. *Ta'rikh as-Sūdān* (French trans. and Arabic text). Paris: Adrien-Maisonneuve (original edn 1913–14).

Hunwick, J. O. 1964. Religion and state in the Songhay Empire, 1464–1591. In *Islam in tropical Africa*, I. M. Lewis (ed.), 296–315. London: Oxford University Press.

Hunwick, J. O. 1969. Studies in the Ta'rikh el-Fettach: its authors and textual history. *Res. Bull. Center Arabic Document.* **5**, no. 1–2, 57–65. December. Ibadan University.

Hunwick, J. O. 1970. The term 'Zanj' and its derivatives in West African chronicle. In *Language and history in Africa*, D. Dalby (ed.), 105–6. London: Cass.

Kaba, L. 1978. Les chroniqueurs musulmans et Sonni ou un aperçu de l'Islam et de la politique au Songhay au 15e siècle. *Bull. Inst. Fondamental d'Afrique Noir* **40B**, 49–65.

Kaba, L. 1981. Archers, musketeers and mosquitoes: the Moroccan invasion of the Sudan and the Songhay resistance (1591–1612). *J. Afr. Hist.* **22**, 457–75.

Khaldūn, I. 1974. *The Muqaddimah* (trans. F. Rosenthal). Princeton: Princeton University Press.

Levtzion, N. 1971. A seventeenth-century chronicle by Ibn al-Mukhtar: a critical study of *Ta'rikh al-Fattash. Bull. Sch. Oriental Afr. Studs Univ. Lond.* **34**, 571–93.

Levtzion, N. 1973. *Ancient Ghana and Mali.* London: Methuen.

Levtzion, N. 1977. The Western Maghrib and Sudan. In *The Cambridge history of Africa*, J. D. Fage and R. Oliver (eds), 331–462. London: University of London Press.

Lewis, B. 1971. *Race and color in Islam.* New York: Harper & Row.

Ly, M. 1972. Quelques remarques sur le *Ta'rikh el-Fettach. Bull. Inst. Fondamental d'Afrique Noir* **34**, 471–93.

Mauny, R. 1961. *Tableau géographique de l'ouest Africain au Moyen Âge.* Dakar: Institute Fondamental d'Afrique Noir.

MBaye, E.-H. R. 1972. Un aperçu de l'Islam ou réponses d'al-Maghili aux questions posées par l'Empereur Askia Mohammed. *Bull. Inst. Fondamental d'Afrique Noir* **2B**, 237–67.

Meillassoux, C. 1975. Captifs ruraux et esclaves impériaux du Songhay. In *L'esclavage en Afrique précoloniale.* Paris: Maspéro.

Niane, D. T. 1975. *Recherches sur l'empire du Mali au Moyen Âge.* Paris: Préscence Africaine.

al-Omari, I. F. A. 1927. *Masalik al-absar fi mamalik el amsar* (trans. M. Gaudefroy-Demombynes). Paris: Geuthner.

Rodinson, M. 1966. *Islam et le capitalisme*. Paris: Semil.
Rodinson, M. 1968. *Islam et le socialisme*. Paris: Semil.
Rosenthall, E. I. J. 1968. *Political thought in medieval Islam*. Cambridge: Cambridge University Press.
Rouch, J. 1953. *Contribution à l'histoire des Songhay*, 184. Dakar: Institute Fondamental d'Afrique Noir.

Saad, E. N. 1979. *Social history of Timbuktu 1400–1900: the role of Muslim scholars and notables*. PhD thesis, Northwestern University.
de Sardan, J. P. O. 1975. Captifs ruraux et esclaves imperiaux du Songhay. In *L'esclavage en Afrique précoloniale,* C. Meillassoux (ed.), 99–134. Paris: Maspéro.
Suret-Canale, J. 1961. *Afrique noire*. Paris: Editions Sociale.

Terray, E. 1975. Classes and class consciousness in the Abran kingdom of Gyaman. In *Marxist analyses and social anthropology,* M. Bloch (ed.), 91. New York: Wiley.
Tymowski, M. 1967. Le Niger, voie de communication des grands états du Soudan occidental jusqu'à la fin du 16e siècle. *Africana Bull.* **6**, 73–95.
Tymowski, M. 1970. Les domaines des princes du Songhay (Soudan occidental): comparaison avec la grande propriété foncière en Europe au début de l'époque médiévale. *Ann. Econ. Soc. Culture* **25**, 1637–58.

Watt, M. 1971. *Muhammad, prophet and statesman*. London: Oxford University Press.

Zikria, N. A. 1958. *Les principes de l'Islam et la démocratie*. Paris: Nouvelles Editions Latines.

3 *Life before the drought: a human ecological perspective*

EARL P. SCOTT

Throughout the Savanna–Sahel zones nomadic and sedentary groups have developed adaptive mechanisms that permit their survival in an uncertain and inhospitable environment (see Chs 1 & 6 for discussions of other adaptive mechanisms). These adaptive mechanisms range from the use of microenvironments to grow staple grains to systematic migrations to obtain grass and water for cattle. The adaptive mechanisms commonly employed by specific ethnic groups in different parts of the Savanna–Sahel zones vary, but some are common among all groups. One such mechanism is cooperation, which involves several forms of exchange between farmers and nomads. These exchanges developed over many centuries and now play a critical role in the survival of the Savanna–Sahel peoples and the ecosystems in which they live. The long established cooperative relationship between the Habe and the Fulbe in northern Nigeria, for example, has been called symbiosis. It began centuries ago with the coming of the first Fulbe pastoralists to northern Nigeria. The exchange relations between these groups developed into a close association that should be described as protocooperation. Indeed, under some conditions and situations it may best be described as mutualism. That is, viewing the Habe–Fulbe as a 'new variational ethnic' group rather than two distinct, competing ethnic groups.

This chapter focuses on the varied relationships between the Habe and the Fulbe peoples of northern Nigeria. The Habe (Hausa agriculturalists) and the Fulbe (Fulani pastoralists) interact with each other and with their environment which they perceive as a store of resources. Spatial and cultural problems arise when, in selecting certain resources to pursue their basic livelihood activities, these groups are brought into proximity. The relationship that has developed as a result of their interaction over time in the same environment but employing different resource-use patterns is described here. The objective is to suggest how a human ecological approach can be useful in understanding how man has learned to live in a very harsh and uncertain environment and to propose that cooperation between nomadic and sedentary groups played a much greater role in the settlement of the Savanna–Sahel zones than would be expected by the constantly reiterated stereotype of internal competition and conflict between them.

The human ecological approach

The human ecological approach to resource use allows us to perceive man as a human organism in nature interacting with other species of organisms (i.e. the nature of the interrelationships between living organisms and their physical and biotic environment). The human ecological approach emphasizes interspecies as well as intraspecies transactions. The ecologists' explanation of community processes and the principle of the balance of nature – i.e. the tendency of the organic community toward homeostasis within the limitations of its living space – have relevant implications for the behavioral sciences which should be exploited by geographers (see e.g. Berry 1964, p. 3; National Academy of Sciences 1965, p. 28; Stoddart 1967, pp. 530–8). One aspect of community development stressed here is the increasing complexity of societies that interact for mutual benefit, which improves their chances of survival in an uncertain environment (Hawley 1971, p. 11). Cooperation, not competition, leads to greater social complexity and survivability. Allee *et al.* (1949, p. 729), for example, state after Leake (1945, p. 252) that 'the probability of the survival of individual living things or of populations, increases with the degree with which they harmoniously adjust themselves to each other and their environment'.

The human ecological approach to resource use also allows us to consider the role of stress, usually environmental, in developing human adaptations. Two types of stress are identified: continual or regular stress, and uncertain or irregular stress. Continual stress results from the exhaustion of a population's resource base. In such a case, individuals are forced to struggle for survival and as individuals they adapt, exist in abject poverty, or perish. Possibly the best example of a human population abandoning its social organization and reverting to individual instincts of self-preservation because of the exhaustion of its resource base is the IK (Turnbull 1972). On the other hand, uncertain stress, which often results from climatic change such as the annual change in rain amount and distribution, has the effect of stimulating group responses. In this case, groups or societies are forced to interact in such a way as to complement the needs of the other in order to enhance their ability to survive. The Habe and Fulbe of West Africa are representative of social groups that have adapted to the very uncertain environmental conditions of the Savanna–Sahel zones, for although rainfall and other climatic factors vary annually and quite extremely in the Savanna–Sahel zones, there appears to have been considerable socio-economic adaptation to just this environmental variety.

Life before the drought among the peoples in the Savanna–Sahel zones was a mixture of cultural or ethnic difference, social conflict, and economic coexistence. The livelihood activities, though steeped in history and having retained much of their traditional attributes, were not unchanging (Buchanan & Pugh 1955, pp. 102–3). Instead, the livelihood activities practiced by the nomads and the farmers were constantly adjusting to new crops, new cattle management practices, new (commercial) attitudes toward production, changes in environment, and changes in societal needs (cash, for example). The following section begins with a brief discussion of the historical description of the continuing relationships that exist between them. The complex nature of the relationships that existed before the drought suggests that the Habe–Fulbe relation was evolving toward the optimum positive interaction pattern, mutualism.

Habe–Fulbe settlement in the Savanna–Sahel zones

The Habe (Hausa) Empire Hausaland, ecologically located in a relatively dry, grassland area of present-day Nigeria, is not well known historically. The Habe (Hausa) legend of 'Abuyazidu' (Johnston 1967, p. 5) suggests that the indigenous peoples and their culture were altered by successive waves of migrating Berbers between AD 500 and 1500. This peaceful fusion took an unknown period of time to run its course, but one manifestation of it was the emergence of the Habe language. Habe legend also suggests that these Berber migrants brought with them a higher civilization which ultimately advanced the civilization of the indigenous sedentary peoples. Unfortunately, there is little evidence, historically or otherwise, to support the Habe legend. On the other hand, works by noted Arabic scholars, oral myths, tradition, and chronicles, especially the Kano Chronicle (Johnston 1967, p. 9), suggest that the people of Hausaland were fairly sophisticated in the smelting and working of iron for at least 500 years and probably more (Johnston 1967, p. 4). In addition, they had developed agriculture, trade, cities, and a system of government based on chieftainship. M. G. Smith characterized the Habe as follows:

> The Habe (or Hausa) were a sedentary people principally engaged in farming grains, such as millet, sorghum, and maize, with a wide range of subsidiary crops. The Hausa manipulated a comparatively high and varied technology and operated a well-developed economy which, although precapitalistic, is particularly interesting for its range of crafts, institutionalized markets, cowrie and other currencies, and long-distance two-way caravan trade . . . the walled towns and its hamlets, now as in the last century, form the typical unit of territorial administration under the community chief, who is ultimately

> responsible through his superiors to the head of the State. (Smith 1960, p. 4)

Another interesting development characteristic of Hausaland was the territorial organization centered on 'city-states' and administered feudally. The Hausa city-states (called Hausa *Bakwai*) included Daura, Kano, Rano, Katsina, Zazau, Gobir, and Garum Gogas. Although Hausaland later incorporated additional territory (as I shall show later), the city-states remained the core. Seven peripheral states, initially culturally heterogeneous but through assimilation virtually indistinguishable from the original Hausa *Bakwai*, were also politically controlled by Habe rulers. Of these seven states, Zambara and Kebbi, and to a lesser extent Yauri, became most closely identified with the Habe (Johnston 1967, p. 6).

All these states were ruled by Habe chiefs, who wielded great, but not unbounded, power and were administered feudally in that they were obliged, when called upon to do so, to render military service with a stipulated number of armed followers under their command (Johnston 1967, p. 9). The states were also great centers of learning and commerce, and by the 14th and 15th centuries they had established communications with the Kingdom of Bornu to the east, Mali to the west, and were linked, by caravan routes, to Egypt and the Christian Kingdom of Nubia (Johnston 1967, p. 12). Katsina, for example, founded in AD 1100, was famous for its learning and was the first state approached by El-Maghīlī in the late 15th century for the introduction of Islam (Johnston 1967, p. 11). During the 16th and 17th centuries, Katsina was at the center of disputes between Bornu and Songhay. Until the Fulbe uprising at the beginning of the 19th century, Katsina and Daura had been ruled by the Habe, the original people. Since this revolt the ruling class in Katsina Emirate has been formed by the Fulbe, whereas in Daura the Habe have remained in power up to the present day (Luning 1961, p. 4). On the other hand, Kano became a major entrepôt for trans-Saharan commerce and according to Johnston:

> They [new routes] thereby set the city on the road to becoming the commercial and industrial center of the Sudan and laid the foundations for the subsequent growth in the prosperity and importance of the whole of Hausaland. (Johnston 1967, p. 12)

Hausaland was not, then, an isolated, stagnate empire. Hausaland, by the 16th century, had spread over much of what is now Nigeria. Islam, although unevenly accepted by the sedentary populations, was well established, trade was flourishing, and communication had improved significantly. But lack of adherence to Islam, interstate rivalry and warfare, and the decline of Bornu, along with other fragmenting forces, weakened the Habe Empire (Johnston 1967, p. 17). These later forces were to play a significant part in preparing the way for the rise of the Fulbe (Fulani) Empire.

The Fulbe (Fulani) Empire No one knows with any certainty the origin of the Fulbe (Fulani). Nevertheless, there has been no lack of theories, ranging from the tenable to the fantastic (see e.g. Hogben & Kirk-Green 1966, pp. 111–13; Johnston 1967, pp. 17–19). Hogben and Kirk-Green (1966) examined at length most of the theories advanced by various writers and selected L. Tauxier's as the most tenable. Their summary of Tauxier's argument is as follows:

> The Fulbe long ago were a branch of a red hamitic race who inhabited regions in East Africa close to the Masai peoples. He thinks that from East Africa they probably went northwards, up the whole length of Egypt, bearing westward along the edges of the desert until they eventually came up against the foothills of southern Morocco. It was from here, in the north, that they came southwards to Tekrur and the Senegal basin to mix, mainly, with the Serer, but also with the Wolof, thus evolving a mixed pastoral race of Tucolor. The suggestion is that they were under pressure from the Ummayad Muslims to leave the Maghrib at about the same time as the expedition of 739 to the Sudan, and they may have followed in its wake . . . Here they seem to have remained until about the fourteenth century, when they began to work their way eastward into the countryside between the settled areas. By the sixteenth century large numbers of them were grazing their herds in Massina, where they maintained themselves under their own chiefs, the Dyalo, who in turn acknowledged the authority of the emperor or other rulers of the day, with whom they had little contact. (Hogben & Kirk-Green 1966, pp. 112–13)

If we accept this view, then, several other points must be considered fully to describe the migration and settlement of the Fulbe into Hausaland. First, because of the cultural and physical differences between the Fulbe and the indigenous people of Hausaland and other sedentary peoples of West Africa generally, their origin is postulated to be outside of Africa. Consequently, Tauxier suggests the original home of the Fulbe is Asia, not Africa (see Hogben & Kirk-Green 1966, p. 113). Secondly, because of the particular nature of the resource-use pattern adopted by the Fulbe, their eastward movement across the Sudan was gradual and irregular. Three types of movements were important in the dissemination and colonization of the pastoral Fulbe throughout the western Savanna–Sahel zones: transhumance, migratory drift, and migration proper. Of the three, migratory drift, which is the continuous adjustment of transhumance patterns to subtle changes of an ecological nature (Stenning 1959, p. 22), was more important. Thirdly, because of competition for the same resource (i.e. grass and water for cattle) by other pastoral peoples moving west (Johnston 1967, p. 81), the Fulbe were not only forced to stop their eastward migration but were diverted south into Bornu and Adamawa of present-day northern Nigeria and northern Cameroon. Today, northern Cameroon contains one of the largest concentrations of Fulbe, who speak Fulfulde, that exists anywhere in Africa (Johnston 1967, p. 82).

As the Fulbe migrated eastward they began to settle among the Habe and other sedentary agriculturalists. Why the Fulbe settled initially seems to be a point of little agreement in the literature.[1] Three points of view are advanced to elucidate this aspect of the Habe–Fulbe relationship. First, Hopen's view as to why the Fulbe began settling emphasizes the loss of cattle by the Fulbe and the subsequent 'order' and British 'justice' imposed after the British gained military control of northern Nigeria. Hopen argues that,

> Having had their herds reduced by the rinderpest epidemic (1887–91) (see St. Croix, 1945, p. 13) and, under British rule, having their slaves emancipated, the Fulbe were in a difficult economic position. It was partly on account of the humiliation of living in a state of poverty in the same village as their newly emancipated slaves, that some of the Fulbe chose to move away from their traditional sites . . . At their home village they began to farm alongside their emancipated slaves, gaining only a fraction of their subsistence from their cattle. Some were able to increase their herds and leave in search of better pasture to the south. Others, whose herds did not increase or whose cattle died, remained at their traditional village. (Hopen 1958, p. 49)

Stenning, writing in 1959 (a year after Hopen), presents a slightly different argument to explain the circumstances influencing the Fulbe's decision to become sedentary agriculturalists. He concedes

> that semi-sedentarism arises principally through losses of cattle by disease, when widespread reductions in the size of herds below the level necessary for entire subsistence upon them to render inoperable the mechanisms of loan or gift which enable the pastoral Fulani herd-owner to recoup his losses. (Stenning 1959, p. 7)

On the other hand, as he explains further, semi-sedentarism may not necessarily

> be the result of poverty in cattle. For example, pastoral Fulani moved onto the Jos Plateau in Northern Nigeria as recently as 1910. They

> found there a high, fly-free grazing ground with abundant water and pasture. Their seasonal movements decreased in scope and their herds multiplied. The growth of the tinning industry and the establishment of creameries assured them profitable markets for their dairy surpluses. At the same time the Pagan inhabitants increased their agricultural holding . . . [making] the need to establish permanent rights to wet-season pasture . . . more desirable. The pastoral Fulani surrounded their settlements with gardens of Indian corn which are cleared, planted, tended and harvested by Pagan labourers paid in cash . . . On the Jos Plateau, it is the richest cattle-owners with the largest families who adopt this form of settlement . . . Semi-sedentarism is here correlated, not with poverty in cattle, but its converse. (Stenning 1959, pp. 8–9)

F. W. de St Croix, who was the first to describe, in detail, the cause and extent of the rinderpest outbreak in northern Nigeria, and who subsequently helped in eradicating the disease from the area (de St Croix 1945, p. 13), did not see the loss of cattle as the main cause of the Fulbe becoming sedentary farmers. In fact, he presents a totally different point of view:

> As regards the relationship of the nomadic Fulani with the (Habe) rulers, it would appear that the majority of, if not all, such tribes had, from long ago, *representatives of their own in the towns of the ruling chiefs*. [my emphasis] It is reasonable to agree with the Fulani assumption that this representation came about in such manner as that, as a young man, a member of a family standing, perhaps the son of an 'Ardo' [Fulani chief or leader], who decided to give up the nomadic life for scholarship and, having been given a wife, retired with her to a town. [Often relatives would visit and stay with him, and exchange news; when necessary, he would perform certain duties for which he was paid a 'voluntary fee'.] These dealings would not go on without the knowledge of the local ruler, who would receive his due portion from this man: moreover he would realise the value to him of one who knew of the movements of these nomadic Fulani . . . thus the scholar would in many cases become the go-between in the rulers' dealings with his nomadic relatives: and, in course of time, an intermediary between the ruler and the nomadic Fulani frequenting the province. (de St Croix 1945, p. 7)

Although this argument does not have as wide acceptance as the loss of cattle argument, the obvious benefits to both Fulbe nomads and Habe agriculturalists alike suggests that it should be considered seriously. Moreover, the representative system described by de St Croix is very similar to the *mai gida* (or landlord) system described by Hill many years later (Hill 1966, p. 3). If it is in fact a parallel system of exchange that has developed outside the marketplace *per se*, its economic and geographic implications become even more intriguing.

Regardless of the view one accepts, the question of whether the Fulbe are in transition between nomadism and sedentarism inevitably arises. Stenning, who addressed this question, feels that the Fulbe are evolving toward a more sedentary, cattleless way of life and that the 'Fulani communities of the western Sudan . . . demonstrate a further stage in the progress towards a sedentary way of life' (Stenning 1959, p. 9) than the Fulbe communities in the eastern Sudan. Moreover, he suggests that where this process is substantially advanced, it 'is often marked by the abandonment, in the home area, of one of the traditional types of pastoral Fulbe shelters in favour of the hut type common among the sedentary populations'. (The spatial implications of these statements have not been investigated.) There is evidence that pastoral Fulbe, who engage in agricultural pursuits as a consequence of cattle losses, do succeed in re-establishing herds capable of supporting them completely, and then take up the nomadic life once more (e.g. Hopen 1958, pp. 1–5; Hogben & Kirk-Green 1966, p. 425). I suggest, however, as Stenning does, that short-term studies of semi-sedentary communities would give the impression that the process of 'reverting' back to nomadism is more widespread than it is (Stenning 1959, p. 8).

Therefore, the evidence would suggest that the Fulbe are becoming more sedentary, thereby contributing to the increasing reduction of pastureland resulting from agricultural expansion, and that some groups or areas in the Savanna–Sahel zones are more advanced than others. Today there are at least three categories of Fulbe groups that can be distinguished based on their livelihood activities. These are: (a) *Bororo'en*, who are characterized as true nomads and are thought of as having vast herds of cattle, being less tied to a particular area and knowing much more about cattle medicine. They favor large tracts of bush, they speak only Fulfulde, and they are 'unbelievers'. (b) *Fulbe na'l* (cattle Fulani) are the semi-sedentary pastoralists. This way of life has many variants, but its essential feature is that the family is no longer completely footloose but has acquired a home base of some kind and engages in farming as well as raising stock; however, half the herd and household is split for half the year while the search for pastureland and water is carried out – usually by the unmarried son. (c) *Fulbe Sire* (town Fulani) include both the aristocratic families such as the Torobe (the Fulani who belong to the ruling and professional classes; their real interests lie in administration, law, religion, and education) and the small peasant types of farmers who no longer keep cattle in any numbers (see e.g. Stenning 1959, p. 7; Smith 1960, p. 5; Hogben & Kirk-Green 1966, p. 110; Johnston 1967, pp. 21–3).

The Fulbe were probably *bororo'en* when they moved into West Africa and for many reasons, some of which are listed above, they gradually, even reluctantly, began to settle as farmers. However, the greatest move toward sedentarism probably occurred after the Holy War (*Jihād*). As I stated above, Islam was introduced into northern Nigeria very early and by the 19th century two major sects, the Kadiriyya, founded in the 12th century, and the Tijjaniyya, founded in the early 19th century, were well established. The acceptance of Islam was by no means ubiquitous in Nigeria or in West Africa generally. Yet Pagans, Habe Moslems and non-Moslems, Fulbe Moslems and non-Moslem Fulbe lived for decades, indeed centuries, in relatively close proximity. It was not until 1804 (although the war actually started in 1725 in the Futa Jallon with the Fulbe and Tucolor and later spread to the Futa Toro with Abd el Kader's revolution of 1776 after the battle of Kwotto Lake) that Shehu Dan Fodio and his son declared a *Jihād* against the enemies of Islam (back-sliders) and, in the next decade or so, Shehu, or his son and successor, Bello, gave the flag of Holy War to trusted followers of emirs who took existing Habe kingdoms by insurrection or carved out new ones by war, all in the name of religious reform (see e.g. Stenning 1959, p. 15; Hogben & Kirk-Green 1966, p. 114; Johnston 1967, p. 24).

By 1810 the Fulani Empire was established and 'the Fulani rulers of the Hausa City States progressively adopted the sedentary habits of the subject (Hausa) population, together with their language (Hausa) and other cultural elements' (Smith 1960, p. 5). These settled Fulani, who were basically administrators and scholars, became known as the town Fulani or *Fulbe Sire*. 'There was some inter-marriage among [these] professional classes with concubines taken from the Hausa community and those who had lost their cattle may have had to seek their wives in the same quarters' (Johnston 1967, p. 25). Islam became more entrenched. Political and military leadership shifted from the Hausa to the Fulani conquerors, but the territorial arrangement and the political, hierarchical structure generally remained as they were prior to the Fulani conquest (Smith 1967, pp. 95–6). In general, it seems that acculturation and assimilation progressed unbelievably smoothly. The probable explanation for this is that the change of dynasty seems to have made little difference to the populace (*talakawa*). Hogben and Kirk-Green note that 'in local Katsina lore, the difference (between Fulani and Hausa rule) is sometimes expressed by simply noting that Koran (Fulani) was *Ja*, "red", whereas Sanau (Hausa) was *baki*, "black"' (Hogben & Kirk-Green 1966, p. 159). In addition, the Fulani probably were not a large enough force, population-wise, to have an effect on the common man, nor did they have as their conquest motive the absorption of other peoples' culture into theirs. The ultimate result of the *Jihād* in northern Nigeria, with notable exceptions such as Adamawa, Muri, and Gombe emirates where Fulfulde continues to be the language of daily intercourse (Hogben & Kirk-Green 1966, p. 429), was the loss of their ethnic language,

Fulfulde, and the adoption of the indigenous language, Hausa. In this respect, the Fulani won the war but lost their culture.

The rise of the Fulani Empire through the activities of the Fulani Holy (learned) men thus depended in large measure on their influence among communities of their own people. But to describe the Holy War as a demonstration of Fulani 'nationalism', in Stenning's view, is not entirely correct (Stenning 1959, p. 23). When the Fulani reformists deemed it necessary they ousted Pagan Fulani dynasties (e.g. in Fouta Jollon, Fouta Senegal, and Macina). In Hausaland some Fulani opposed Shehu and even fought on the side of the Hausa Kingdom, whereas on the other hand, Shehu gave flags to Hausa and Bornu preachers as well as to Fulani (Stenning 1959, p. 23; Hodgkin 1960, p. 40). Finally, as I mentioned above, Moslem Fulani themselves belonged to rival sects that were well established by 1804. Although nationalism did not assert itself at this time, Hausaland, which was initially fragmented politically, was consolidated under a central political administration.

The Fulani 'inherited' the feudal system of the conquered Hausa, and perpetuated it with little change. At the point of conquest the emirs acquired considerable but not unbounded power. And their positions were supported by a system of taxation. These taxes fell into four broad categories: (a) *haraji* or *kardin kasa*, which was a poll tax or general tax on the farming community; (b) *jangali*, the tax on cattle; (c) death duties, which were levied according to Moslem law; and (d) there was a miscellaneous group of impositions on crafts and trade (Smith 1960, pp. 142–240, & 1967, pp. 103–4). Taxation made all the various peoples subject to the emirs. In addition, the establishment of a single government over Hausaland for religious and/or economic reasons provided a reasonable degree of internal security and cohesion in a region that had formerly been controlled by petty warring states and stimulated a considerable amount of internal and foreign trade (Hodgkin 1960, p. 42). Islam, more than any other factor, homogenized these groups by imposing a common life-style that they could adopt, if only partially by some.

It is within this historical context that the Fulani–Hausa relations developed to where they are today. The next section of this chapter deals with the variety and complexity of these relations. They will be discussed under four broad headings: (a) environmental relations; (b) social relations; (c) political relations; and (d) economic relations. The underlying theme presented here is that the Hausa–Fulani, though historically distinct, are now highly interdependent, to the point of mutual necessity.

Hausa–Fulani relations

Environmental relations The Hausa–Fulani occupy the ecological area of northern Nigeria called the Savanna–Sahel zones. The territorial organization in this relatively dry, easily traversed grassland emphasizes defensive values and is based on the compact distribution of population within walled towns and nucleated villages, usually along trade routes, and administered by a local community chief. 'Each of these towns had a few smaller settlements near it which owed allegiance to the village chief of the area in which they were sited. Many, but not all, of the hamlets were slave-villages' (Smith 1960, p. 91). The Fulani nomads' resource pattern forces them to operate in and among these compact, nucleated settlements. And, as Stenning points out, 'pastoral Fulani did not, and do not, merely graze at will, but obtain rights to the facilities they required from the acknowledged owners of the land . . . Rights to graze were affirmed by the payment of tribute to the ruler or his local representative' (Stenning 1959, p. 6; also see Johnston 1967, p. 25). Given the nature of these resource-utilization patterns (i.e. Hausa, intensive cultivation; and Fulani, extensive grazing), benefits accrue to both parties when they cooperate. But conflicts inevitably arise over land, water, and rights of way.

Possibly the best example of conflict over resource use exists between the pastoral Fulani and the sedentary peoples of Gwandu. The problem centers on the use of the Sokoto, and to a lesser extent, the Niger floodplains. In these river valleys ecological conditions are favorable, not only for cattle grazing and farming, both of which can be done year round, and especially so in the dry season, but sedentary fishermen also gain their livelihood from catching and marketing fish over a wide area of the western and eastern Savanna–Sahel zones (Hopen 1958, p. 20). The subsequent conflicts that arise as a result of people being forced into a given region because of common resource-use interests, although they are separate and distinct, are discussed by Hopen. He observed:

> Because of their common interest in the floodplain it is not surprising to find the greatest concentration of pastoralists and farmers within easy reach of the Sokoto River in particular. However, the effect of farmers cultivating both flood-lands and upland from a single home means that the most intensive upland cultivation is also found near the banks of the Sokoto River . . . now it has been said that pastoral movements are southward in the dry season and northward in the wet season. But they must also move into the flood-plains in the dry seasons and into the upland bush in the wet season. Thus in order to meet their ecological needs herdsmen must have seasonally appropriate access to both the flood-land and the upland bush. But without enclosure and without an adequate number of well-demarcated and recognized cattle trails it is not possible, in these circumstances, to guarantee the protection of the interests of both pastoralists and farmers as they each, in their own way, go about gaining their livelihood from the land. (Hopen 1958, p. 22)

The problem of land allocation and rights of way has been alleviated, at least in some areas, by mutual agreements.[2] Stenning reports that 'for guaranteeing cattle tracks and the use of water supplies' the Fulani must graze their cattle on certain areas of farmland (Stenning 1959, p. 6). But because the Fulani tend to move frequently, their right to land for any purpose is tenuous to say the least. Furthermore, Fulani herdsmen seek out, move to, and possibly remain in those areas ecologically suitable to their cattle. It seems reasonable to say that ecological conditions influence the size and distribution of local cattle grouping and, as we shall see, have an influence upon the nature of the relationship that exists between the Fulani and the Hausa. In this aspect of the Fulani–Hausa relationship, cooperation and conflict are inevitable.

Social relations When we consider social relations between the Fulani and the Hausa, the status of the Fulani as a minority group becomes more apparent. Their population numbers did not allow them to impose every aspect of their culture on the conquered Hausa. They depended to a large extent on non-Fulani people to fill these positions. Stenning states:

> The latter [non-Fulani] were part of the Fulani social order at all levels. They had their place in Fulani ceremonial, mastered Fulani etiquette, and spoke the Fulani language, and many of their members might claim to be 'part Fulani.' [He notes also that] in spite of the decrease of ordinances of colonial governments against various forms of slavery, these relationships persist, although with many modifications. Communities of slave or serf origin may have the outward appearance of Fulani communities of sedentary farmers, semi-sedentarists, or even pastoralists. [He suggests that] their origin and present status require careful elucidation. (Stenning 1959, pp. 17–18; also see de St Croix 1945, pp. 5–7)

Although many of the villages established by the Fulani rulers were slave villages (de St Croix 1945, p. 5), today there seems to be voluntary residential separation practiced by the Fulani. Smith believes that this is due, in part at least, to a 'clearly conditional and indirect loyalty' the Fulani have for the emirs (Smith 1960, p. 240). Their immediate loyalties are to their heads of lineage. Smith observed that, within their village, Hausa village chiefs

> naturally have the highest individual status, but where semi-nomadic Fulani have settled in Hausa village areas they maintain an exclusive settlement and unity under their own Ardo [head], who deals with the Hausa village chief, and may also deal with the District Head direct. (Smith 1960, p. 251)

More important, this practice seems to be widespread and is very old in northern Nigeria. For example, some Fulani visit the capital cities of Hausaland such as Yola, Kano, or Zaria, where they live, 'not as homeless immigrants ill-adjusted to the life of the city, but in centuries old "quarters" where people of their own regional or ethnic background congregate' (Stenning 1959, p. 22). It is also interesting to note that a special set of arrangements applied to Hausa Moslems (mainly immigrants from other parts of the empire). In addition to being subsidized with grain, labor, and other assistance until the next harvest, they were also assigned special fallow farmland and compounds (Smith 1967, p. 103). These strangers' towns (or *Sabon gari*) are distributed unevenly throughout Hausaland and owe allegiance not to the local chief but to the district head located in one of the capitals. Either because of their religion or because of reciprocal services rendered, these strangers were exempt from local corvée and fines, and their grievances were entertained at the district level, not the local level (Smith 1967, pp. 103–4). Socially, then, the village peoples and the inhabitants of the *Sabon gari* remained separate and distinct. Even in the marketplaces the Fulani and the Hausa seldom interact socially. Fulani men go to the market to socialize, but the Hausa, who live in the village, have less incentive (socially) to visit the market. 'Thus there may be little social contact between the village-dwellers and the herdsmen within the village market area. Moreover, as has been said, when they do visit the villages Fulbe associate mainly with one-another' (Hopen 1958, p. 154). Strict social separation is breaking down now that *sharoo* (Fulani flogging) and other ceremonies are held in Hausa marketplaces.

Possibly the most important factor in describing the way in which Hausa and Fulani interact socially is the way they view each other. For example, the effect of Islam simultaneously links and divides the Hausa and the Fulani within Hausaland. Islam supplies a common and easily identified religion and legal system, a common framework of theory, and a technique for government. Islam also gives religious and moral credence to Fulani rule. Accordingly, Islam has contributed to the Fulani's self-image of guardian and teacher of the Faith on the one hand, and the Hausa as the wards and pupils on the other (Smith 1960, p. 5). But even more important than this is 'the influence of the "time of war" which still actively dominates social relations' in Hausaland (Hopen 1958, p. 147). The Fulani find it difficult to forget the social status the Hausa once held (i.e. their servile status), and when conflicts arise over land for pasture or cultivation, these former differences between them are revived and perpetuated. Hopen observed that

> in their relations with the Habe the herdsmen see themselves as a group who are not only ethnically and culturally distinct [and superior], but who are also occupationally specialized and have a common body of interests and problems. (Hopen 1958, p. 147)

The Hausa, on the other hand, view the Fulani as fanatic cattle lovers who, in consequence, have little more 'intelligence' than their animals. Furthermore, the Hausa feel very justified in contesting the claim of the Fulani to be the local elite by asserting that they themselves are today more learned and orthodox in Islam (Hopen 1958, p. 148).

When we consider residential location, group identity, perpetuation of past attitudes, aloofness of the Fulani, and even fear of cultural and physical annihilation through assimilation and acculturation (Hopen 1958, p. 27), it becomes clear that centuries of living in close proximity has not resulted in an indistinguishable social system in Hausaland. Clearly the nature of the relationship that binds them together also preserves their distinctiveness. In their minds they perceive themselves as Hausa and Fulani. In reality they exist as Hausa–Fulani (i.e. one ethnic group).

Political relations By 1810 the *Jihād* in northern Nigeria was just about complete. Some of the consequences of the Fulani revolution have been discussed. To put them in perspective, they should be reiterated here. As was stated above, the most obvious consequence of the Fulani 'missionary impulse' was the establishment of a single government, based on Islam, over all of Hausaland. But other aspects of life in Hausaland were also affected. For example, as Islam became more entrenched, the new political system provided a reasonable degree of internal security, internal and foreign trade expanded, and learning and literature were stimulated. Although these changes obviously occurred, it must be kept in mind that the Fulani were, and still are, a minority group and the takeover affected the ruling elite class to a greater degree than the common man (*Talakawa*) in Hausaland. Consequently, the Fulani inherited and perpetuated, with few exceptions, the Hausa system of government, administration, and territorial organization. In other words, changes were made when it was deemed necessary to bring the government more into line with Islamic norms as conceived by the Fulani reformers. Therefore several offices were established. Of these, the office of *Sa'i* was by far the most important. It was charged with the responsibility to 'administer the nomad Fulani, to arrange for their help in war, to settle their disputes, and to collect the cattle tax' (Smith 1960, p. 142).

Significantly enough this ideology of governmental organization, oriented toward the maintenance and expansion of the Muhammadan religion, initially weakened the opposition of the Muhammadan Hausa to Fulani rule. 'Moreover', as Smith so vividly states it,

> as free Muhammadans, the Fulani and Habe were committed by their common interests to try to maintain the current social and political order; numerically also they were together strong enough to prevent a successful revolt of their internally divided slave populations. Within

> the system of competing patrilineages, clientage served to bring Fulani and Habe into close political association, thereby reducing the separateness and unity of these conquered groups. Solidary political relations of clientage were the usual basis of Habe appointments to office (though usually subordinate to the Fulani noble lineage); they provided their holders with opportunities for the accumulation of wealth, for upward social mobility, and for the exercise of power. (Smith 1960, p. 88)

Clearly, these integrative elements operated initially among the wealthy governmental elite and the attitudes developed there presumably filtered down to the common man in Hausaland. Also these political relations seem to have developed in an unstructured trial-and-error atmosphere, the nature and consequences of which were totally unknown to either the conquerors or to the conquered. But the Hausa and the Fulani did merge, although accidentally, under a political system devised by the Hausa and administered by the Fulani. The Hausa system of titled offices acquired through hereditary rights illustrates this point best. Apparently, the practice was widespread throughout Hausaland (Smith 1960, pp. 103–4), but perhaps 'the proliferation of hereditary rights went further in the Sultanate of Sokoto than in other parts of the Empire' (Johnston 1967, p. 113). The significant point is that the Fulani rulers appointed Hausas as well as Fulanis to fill these offices. The 'Chief Justice', for example, 'became a monopoly and was enjoyed by the family of Mallam Mustafa, whom Bello had appointed in 1817, right down to the year 1898' (Johnston 1967, p. 172). On the other hand, the collection of occupational taxes (usually from craftsmen, but entertainers and brokers were also included) was given to the heads of the craft guilds, who were almost invariably Hausa and were recognized by the Emir (Fulani) as *Sarkin Makera* and *Sarkin Fawa* (both well established Hausa titles) (Johnston 1967, p. 173).

At a lower level of government other relations developed that were integrative and contributed much toward present-day political relations in Hausaland. Four titled offices were involved: (a) *hakimai* (resident chiefs); (b) *masugari* (owners of the village); (c) *Sarkin Fulani* (spokesmen for the Fulani nomads); (d) *rukuni* (administrators of the *Sabon gari*).

Briefly, the indigenous population of Maradi was administered by resident chiefs (*hakimai*), each of whom controlled several contiguous villages under local headmen, *musugari* (owners of the village). (These offices carried a degree of prestige, were avenues to wealth and power, and were also obtained through hereditary rights.) Among these sedentary peoples, nomadic Fulani grazing their cattle inevitably raised conflicts. The *Sarkin Fulani* mediated between the pastoralists and the settled farming population and, with the local *hakimai*'s approval and council, investigated disputes, supervised grazing rights, collected the annual cattle tax, settled civil issues of divorce, inheritance or debt, and carried out many other duties for which he received a fee, a portion of which went to the local *hakimai*. Moreover, the *Sarkin Fulani* was 'especially required to report all Fulani movements into or out of the territory, and to patrol the cattle routes when required' (Smith 1967, p. 103). The need for cooperation and communication between the *Sarkin Fulani* and the *hakimai* is obvious, if for no other reason than their mutual economic and political interests.

On the other hand, the migrating Hausa Muslims created another problem which also required clever political manipulation involving both Hausa and Fulani. The Fulani emirs were forced (by religious affiliation) to deal with this category of Hausa differently than the others (i.e. pagans). When these migrants settled in a village administered by the local *hakimai*, the *rukuni*, who lived in the capital and was Hausa, was notified and the immigrants were brought to him. At this meeting the newcomers would pledge their allegiance to the *rukuni* and he in turn would agree to ensure their protection. The *rukuni* also saw to it that the immigrants were fed and fallow farmland was allocated to them. These immigrant settlements became known as strangers' towns (*Sabon gari*). The *Sabon gari* were linked socially, economically, and politically to an individual *rukuni* despite the fact that they were dispersed throughout Hausaland. Smith's insightful comments on this point not only identifies the area of conflict but recommends this aspect of the Hausa–Fulani relationship for further research. He states:

> [in consequence,] *hakimai* normally administered areas which contained a number of Muslim Hausa subjects of different *rukuni* over whom they exercised no jurisdiction. Such Hausa were directly responsible to their Muslim patrons at the capital. To these they paid their tax and took their complaints or requests; from them they received instructions and orders. The relations thus instituted were not personal: the immigrants' issue remained and still remains under the jurisdiction of the *rukuni* title. (Smith 1967, p. 104)

A great deal of power was generated by this system and, as we have seen, much of it was vested in the hands of the *rukuni*. Clearly, the *rukuni* did not always represent the interest of the *Sabon gari* inhabitants and in some cases he probably lost their support completely. However, with their staff, their *Sabon gari* subjects, their supervisory role over *hakimai*, etc., the *rukuni* were, if united, probably more powerful than the Fulani chief. Moreover, in Maradi at least, the main military force consisted of the cavalry and the infantry. The cavalry was recruited from the 'nobles and their slave staff', but the 'infantry and the bowmen were recruited as needed through the *hakimai*' (Smith 1967, p. 119). So, it seems that mutual needs and benefits required the cooperation of both ethnic groups, and although many of the holders of these titled offices today can trace their ancestry back to one or the other group, the system operated in reality as a single, homogeneous ethnic group, the Hausa–Fulani.

Admittedly, the levels of political interactions described above affected only a small percentage of the total population: those who occupied high positions in the political hierarchy. Lower echelon political positions, especially those involving village administration, were also filled by appointment. Smith observes, however, that the Fulani method of political appointment to village posts was characterized by its departure from the Hausa principles of appointment to office (Smith 1960, p. 142). He also states that 'Musa is credited with the order that *hakimai* should remain at the capital, thus separating them residentially from their fiefs' (Smith 1960, p. 143). He suggests several reasons why this decision was made, but fails to say who or from what group of administrators Hausa were recruited to fill this position at the village level. Stenning and Hopen have addressed themselves to this question (Hopen 1958, Stenning 1959), but their discussions focus on a period after Nigeria became a protectorate. Consequently, I am unable to account for a considerable period of village administration. This section, then, will deal with the effects of protectorate status on the political relations of the Hausa and the Fulani. It is significant to note here that the Fulani were affected most, which in turn affected their relationship with the Hausa.

The most obvious consequences of Nigeria becoming a protectorate were the enforcement of quarantines of diseased herds (de St Croix 1945, pp. 11–13) and a more systematic collection of the cattle tax. One would have thought that the movement of cattle would have been decreased substantially: 'on the contrary, the establishment of the protectorate has widened the choice of pasture by opening up highland areas which were formerly denied to the pastoralists by the intransigence of the inhabitants' (Stenning 1959, p. 208). In other words, imposed security has substantially reduced the political, if not the ecological, hazards of migration and thereby increased the range of the nomadic Fulani. (One side effect of this has been to increase the Fulani's dependency on market relations that they carry on with the Hausa. More will be said about this in the next section.)

Village heads were, and still are, chosen from the Hausa, but never, in the past, from the Fulani. Apparently, after Nigeria became a protectorate, the British deemed it necessary to make Fulani village heads also. And, according to Stenning, they were chosen from at least five categories of Fulani: (a) the chiefs who obtained village headships and their retinues; (b) pastoralists who are of the same lineage groups as the village heads; (c) clansmen of the village head; (d) Wodaabe whose lineage or clan is not represented in any village headship; and (e) non-Wodaabe immigrants (Stenning 1959, p. 224). Of the five categories, (a) and (b) were by far the most important in affecting the political relationship between the Hausa and the Fulani.

The effect of the establishment of village headships upon the chiefs and their retinues is straightforward and perhaps, depending on your point of view, unfortunate. The chiefs were chosen initially because of their wealth in cattle and their position in the core-lineage of a clan chiefdom. So, when they took office they had cattle. 'But', according to Stenning,

> by the terms of their establishment as village heads, they were prohibited from seasonal movement and, even more, migrations were forbidden them. [Their cattle had to be disposed of] usually as gifts to their followers or they were sold to meet the various expenses incurred in maintaining their position. (Stenning 1959, p. 225)

For our discussion, this not only suggests that these chiefs have lost their cattle, but that they have become more sedentary, have lost their Fulani identity, developed interests foreign to their pastoralist followers, mastered Kanuri or Hausa, and in general have come to have more contact socially, economically, and politically with the sedentary peoples of the village than the Fulani pastoralists who may be in their village area for only a short period. In other words, 'the Fulani village heads and their retinues are more closely akin to the sedentary villagers' (Stenning 1959, p. 225) than the Fulani pastoralists. It is also significant to note, as Stenning argues, that these chiefs play a unique role during the wet season for their pastoral kin. That is,

> They appear to some extent in guise of functionaries concerned with the transmission of the ordinances of the Native and Provincial administration, particularly those to do with cattle-tax . . . the village head is thoroughly conversant with pastoral conditions in his village area and is consulted by his following . . . he is at this season the arbiter of relations between the pastoralists and the Native administration. (Stenning 1959, p. 225). (It should be noted that de St Croix placed this development, which he called the scholar-representative, much further back in history.)

On the other hand, the effect of the establishment of the Fulani village headships has perhaps been greatest on those pastoralists of the same lineage groups as the village heads. The evidence is quite clear that these groups have substantially reduced their range of migration. Moreover, they now share in whatever prestige is believed to accrue to their village heads. In addition, some real pastoral advantages are to be had by staying in one area: that is, their detailed knowledge of ecological conditions is advantageous when they must compete with newcomers for scarce resources. Also, because they have established good relations with the villagers, 'they can probably rely more heavily on cooperative herding and loan arrangements' (Stenning 1959, p. 227), especially in the dry season. For these reasons, lineage groups among the Wodaabe have become sedentary. As Stenning observed, 'It is significant that no permanent migration out of Bornu is recorded for these groups since 1918' (Stenning 1959, p. 227).

The Fulani have always come under the authority of the village head in whose areas they are resident. But because the Fulani pastoralists choose to move from one area to another, never really establishing residency in a village, the village head can afford simply to ignore them. Furthermore, 'the pastoralists are not regarded by the Habe as regular members of the village community'. And, more importantly, a Fulani pastoralist 'never has the same sentimental, social, political and economic attachment to a given village area as does a Hausa' (Hopen 1958, p. 149). Therefore, he fails to participate in those decisions that directly affect his way of life.

The political position of the average Fulani pastoralist is obviously weak. Similarly, settled Fulani feel that as a result of the establishment of the village headship office their clan solidarity, as well as their traditional power of chieftaincy, has also been weakened (Hopen 1958, p. 52; Stenning 1959, p. 232). Moreover, those things the Fulani formerly cherished, such as their society based primarily on moral sanctions enforced by the *Ardo*, and their racial and occupational exclusiveness, have been challenged and in some cases superseded by administrational changes that have taken place in the last 40 or 50 years. For the Fulani, assimilation and acculturation seem to be irreversible forces, transforming their life-style into a more sedentary, agricultural one and expanding the need for land for residential and farm use. These recently converted farmers, rather than the well established Hausa farmers, moved into ecologically marginal lands and this may have contributed to their subsequent decline.

Economic relations Thus far we have discussed the environmental, social, and political relations that exist between the Hausa and the Fulani. Economic relations between the Hausa and the Fulani are much more explicit and reflect the mutual needs and benefits that accrue to both through cooperation. For several reasons, basically the scarcity of pastureland and political isolation, large communities of Fulani can no longer exist. The most efficient subsistence unit, under conditions of decreasing pastureland and intolerable conditions in the village, seems to be the family household (Hopen 1958, p. 151). The question arises then: how does this unit of subsistence remain viable? Clearly the Fulani depend on sedentary farmers to some degree for their existence. This coexistence is considered a symbiotic relationship. Hopen suggests that,

> indeed, it is for this reason that such a small unit as the family household can maintain its viability or, more correctly, its

> independence of other pastoral households. Economic interdependence between individual pastoral households and the Habe in general is infinitely greater than between any two pastoral households. (Hopen 1958, pp. 151–2)

This interdependence between nomads and farmers assumes many forms. For example, settled Fulani, who have relatively small herds, will combine them so they may be grazed by one herdsman, enabling the others to engage in other work, usually cultivation (de St Croix 1945, p. 26). In another instance, the semi-sedentary Fulani community split both the family and the herd. The household head remains where the farm lies, entrusting a portion of the herd to his married or unmarried sons (Thompson & Addoff 1958, p. 339). They graze the cattle and return in time to assist in farm clearing and harvest. When the family unit cannot supply its own manpower needs, cooperation from kinsmen is solicited. On occasions 'paid labor is engaged to help in farm work' (Stenning 1959, p. 7; also see de St Croix 1945, p. 15).

By far the most important activity in which the herdsmen come in contact with the Hausa of their village area is when the cattle are brought to manure their fields. The details of the arrangements for this practice vary considerably as does the fee paid to herd owners (de St Croix 1945, p. 26; Hopen 1958, p. 155). In some cases the herd owners manure the same farms year after year, but more often, they vary the farms (and even the village) from year to year. There is no fixed fee. Payments, based on the size of the herd and the length of stay in addition to local demand, may be in cash or kind, usually millet or guinea corn. Naturally, where the towns are large and the demand is great, large herds may be attracted. However, grazing conditions are never good enough to support very large herds, thereby serving as a constraint on manuring (as well as to reduce property damage and inter-ethnic conflicts). Therefore, the season for manuring varies locally with pasture conditions and harvest times.

Exchange associated with the practice of manuring is the most important manifestation of interdependence between nomads and farmers. The nature of the exchanges and the practice of manuring has changed over time but both persist to this day in most parts of the Savanna–Sahel zones. Exchange and manuring continue to be the basic adaptive mechanisms in the Savanna–Sahel zones; thus their nature and complexity are discussed here in detail.

Throughout the Western Sudan, from northern Cameroon to Senegal, traditional livelihood activities persist, although they have been modified by modern, European methods of agriculture, by emphasis on selected crops with export potential, and by production for sale rather than for subsistence. The persistence of traditional livelihood activities and the degree to which they have been modified are revealed best in farming. Under sedentary farming, land is intensively cultivated year after year. Rainfed cultivation is practiced on the uplands where grains – millet and sorghum predominantly – are grown. Irrigation cultivation by shadoof (Hausa, *jigo*) is practiced on the lowlands or *fadama* where vegetables are grown to supplement the family food stock (Haswell 1953, p. 25) and to generate cash income (Scott 1974). Upland and *fadama* cultivation are complementary and represent the major source of food for farmers and herders alike. (For a detailed discussion of *fadama* land, see Ch. 7.)

The most striking feature of traditional agriculture in the Savanna–Sahel zones is its reliance on organic fertilizer. Land surrounding villages and dispersed compounds is heavily manured. Land further away – up to 5 miles from the village or compound – is not manured. Animal dung, fowl dung, and household sweepings, including latrine manure (*tokin masai*) and compost (*juji*) are commonly used. Green manure is seldom used, but certain plants such as the *Acacia albeda* are intentionally grown and protected because of the soil-enriching properties associated with their root systems.

Manuring usually begins just prior to the rains, but some farmers begin carting manure to their fields as early as February. Early manuring is associated with *bizne*, the practice of planting some weeks before the rains in hope of receiving sufficient and sustained moisture to produce a crop that can be harvested weeks before those crops that were planted after the rains were well established. The most common method of manuring is by grazing cattle on crop residue fields, which usually occurs from harvest onwards. Manure is applied to upland and *fadama* fields and is the principal method of maintaining soil fertility. Arable land around Kano, and possibly other densely settled areas, presently under cultivation is 'being kept in condition by plentiful application of organic manure' (Hill 1966, p. 57).

Manuring, by grazing cattle on crop residue, reflects an ancient association between the cattle herding and the farming societies in the western Savanna–Sahel zones. Field manuring, however, often leads to intra-ethnic conflict. Indeed, some herding groups, who move into a territory late, often compete with established herding groups for grazing rights on farmers' land, with whom they in turn enjoy a cooperative relationship and are thereby 'cementing the territorial community' (Hill 1966, p. 64). Y. B. Usman is quoted as saying, 'the commonly expressed view that pastoralists and cultivators are in perpetual conflict is an oversimplification: in the Kano close settle zone the farmers have a particular need for cattle manure, and the pastoralists require the highly nutritious fodder of the corn stubble' (Hill 1966, pp. 61–5). The mutual benefits derived through cooperation are substantial.

Crop yields, especially grains, are surely improved by the application of manure to the soil. Haswell observed that 'controlled use of livestock seems to make possible better yields of millet. Millet yields in the Toranko village of Jollof on the borders of the Genieri millet lands are generally higher than the millet yields of Genieri' (Haswell 1953, p. 48). Similarly, Anthony and Johnston observed that 'In this period [after harvest when cattle are grazing on crop residue], the condition of the animals is at its best' (Anthony & Johnston 1958, p. 9). This highly nutritious food period, however, is followed by 6–8 months of very poor conditions and cattle must be moved to wet-season pasturelands. The essential point is that field manuring results in mutual benefits to herders and farmers, benefits they could not achieve by acting alone. In addition, conflict, when it does occur, is usually intra-ethnically related rather than inter-ethnically related.

On the other hand, herders in isolated areas are not influenced by access to stubble. Stenning observed that in West Bornu,

> there is little evidence ... of the symbiosis of agriculturist and pastoralist involving the stubble-grazing and manuring reported for other areas of northern Nigeria. [Instead] Seasonal movements are governed by the search for water supplies rather than by the quest for pasture and the avoidance of, or accommodation upon, farm land. (Stenning 1959, pp. 211–12)

In some areas of former French West Africa, scholars have lamented the 'dissociation of agriculture from animal husbandry' as economically unproductive, where the Fulani refuse to use their cattle as draft animals and where the agriculturists fail to obtain fertilizer, transport, or ploughing from their cattle (Thompson & Addoff 1958, p. 339). In these instances, available resources, especially manure, are not employed to intensify agricultural production, and land has apparently been cropped continuously for several years without 'controlled' manuring. Obviously areas such as these, which lack controlled manuring, are potential food crisis areas.

Although the practice of manuring to improve the fertility of soils varies from one region to the other, cooperation, not conflict, is the overriding basis for interaction between herders and farmers in the Savanna–Sahel zones. In West Bornu, where symbiosis between agriculturalists and pastoralists apparently does not exist, conflicts between farmers and nomads are rare. According to Stenning, this is because 'low sedentary population density ... makes competition and conflict between pastoralists and agriculturalists for the use of land a negligible feature. There is little evidence here (West Bornu) of disputes arising from trespass of herds on farms' (Stenning 1959, pp. 211–12). In essence, field manuring is a significant aspect of cooperation between herders and farmers but, as we have seen, it is only one of several in which they interact for mutual benefit.

This inter-ethnic dependence is very complex and it seems to represent a group, rather than an individual, response to an uncertain and often harsh environment. In addition, the discussion of the various forms of interaction between farmers and herders illustrates their enduring quality, for with the exception of stubble-grazing, few of these patterns of interaction have changed drastically. This is true although there has been a substantial shift to the production of selected crops – groundnuts and cotton – for export to overseas markets.

Other exchange relations Manuring cannot be overemphasized. It is an activity with important sociological implications because it demonstrates the obvious interdependence (or mutual dependence) of two ethnic groups who engage in distinct livelihood systems. Manuring maintains the fertility of the soil which allows permanent, intensive cultivation in an area of inherently infertile soils. On the other hand, manuring provides an income to herd owners without employing additional manpower and substantially aids the family household to function as an economic unit. Hopen suggests that without this help the pastoralists would have to farm on a larger scale and either pay wage-labor or divert the domestic labor force from the tending of cattle (Hopen 1958, p. 155).

Another important aspect of the Hausa–Fulani relationship is the ever-increasing dependency of Fulani on goods made and subsequently sold in the market by Hausa. The Fulani's chief occupation is dairying. They rely on cash income from the sale of milk and milk products to the Hausa in order to purchase their staple food, which is corn (millet) – not milk (Hopen 1958, p. 152). Possibly the most revealing evidence of the Fulani's increasing dependence on the marketplace is seen in Table 3.1. Practically all the items the Fulani consumed were acquired through Hausa markets. In addition to this kind of dependency, Fulani women also developed a relationship with Hausa women which centers on the marketplace. Briefly, Hausa women, who had stalls in the village market, retailed milk brought in by Fulani girls or young women. When the produce had been sold, the cash and the calabash were returned to the girls to be taken home. At the end of a week the older Fulani women would visit the milk retailers and give them a present for their services (de St Croix 1945, p. 30). Today this indirect participation by Fulani women has changed and Fulani women now sell directly to the ultimate consumer.

It may be argued that the mutual dependence of Fulani and Hausa in manuring and other economic situations serves to mediate the hostilities generated between them when Fulani cattle, for example, cause crop damage. I would argue, however, that it is precisely because of conflict and cooperation, both recognized as inevitable, that the two separate societies persist, despite their locational proximity, and do not merge into an indistinguishable entity. This means that under certain conditions and situations they are distinctively Hausa and Fulani; under other conditions and situations they are Hausa–Fulani.

Table 3.1 Market dependency.

Purchased in the market	Produced domestically
household equipment	
calabashes	sleeping mats (men)
mortars and pestles	ropes (men)
water (and cooking) pots	
spoons and ladles	
sleeping platforms	
knives, hoes, and axes	
needles	
fire strikers and flints	
clothing	
cloths	spun cotton (women)
undertrousers	
robes	
turbans	
fezzes and straw hats	
head-cloths	
sandals	
foods	
corn	*gero* (millet)
rice (rare)	milk
meat and fish	butter
spices	meat (only ceremonially consumed)
salt	
delicacies	
kola nuts	
tobacco	
miscellaneous	
medicines, amulets, and charms	medicine, cattle (men)
beads, bangles, rings, etc.	medicine, human (men and women for social and clinical purposes)
raw cotton	
henna, kohl, scent, and soap	
swords, daggers, spears, bows, and arrows (poison)	

Source: Hopen (1958), pp. 152–3.

Rethinking symbiosis

The relationship described above between the Hausa and the Fulani who live in close association is popularly termed symbiosis. They coexist and interact to ensure their mutual existence under conditions of uncertain stress. On this point all of the authors are in essential agreement. For example, Hopen states:

> the Fulbe, living in close symbiosis with the Habe, have a social system which although maintaining some distinctiveness has, nevertheless, been sensitive to political fortunes and changes within the wider Hausa-speaking community. (Hopen 1958, p. 27)

M. G. Smith, describing the political situation at Maradi, states, 'as muslims ruling a pagan population, the Katsinawa of Maradi administered a plural society held together by external threats and internal symbiosis' (Smith 1967, p. 121). Johnston states:

> [the civilization] of the Sokoto Empire was the product of the union between two different strains . . . the two peoples were complementary to one another and between them they evolved a society which was probably more advanced than any other hitherto produced in black Africa. (Johnston 1967, p. 173)

In each of these statements the emphasis is on the living together of two dissimilar groups in close association, the advantages that accrue to both, and the complexity of the subsequently merged societies. The problem raised here is obvious: any positive relationship between two unlike organisms is a symbiotic relationship. Consequently, symbiosis is an ambiguous term that does not precisely describe the relationship between the Hausa and the Fulani.

Ecologically speaking, there are three positive interactions between populations of two or more species, perhaps representing an evolutionary series (Odum 1963, p. 107). They are: (a) commensalism, a simple type of positive interaction in which one population benefits and the other is not affected to any measurable degree; (b) protocooperation, an interaction where two populations benefit each other, but are not essential to each other's survival; and (c) mutualism, an interaction where the association is necessary for the survival of both populations. When these definitions are considered, the evolutionary character of the interactions must be kept in mind. That is, through time an interaction pattern considered to be commensal may evolve into a protocooperational relationship, or even a mutual relationship. Moreover, the suggestion is that mutualism represents the ideal form of positive interaction between two or more species.

The evidence presented above suggests that competition and cooperation exist in the Hausa–Fulani relationship. For example, the Hausa and the Fulani compete for the use of two major resources, water and land. On the other hand, different groups of pastoralists compete for one resource, grassland. Finding a solution to the latter has been approached in two ways: (a) reduction in the size of the viable subsistence unit, and (b) reduction of grazing range, thereby becoming more knowledgeable of local ecological conditions and water supply.

The competition over land is more complex. The problem centers on differences in resource-use patterns: one intensive (farming); the other extensive (grazing). Mixed farming has been suggested as a solution, but it has not been successful (Stenning 1959, pp. 237–47; Scott 1974). Its objective was systematically to make the Fulani pastoralists more sedentary. Some have become farmers, but clearly all Fulani will not become sedentary. Mobility will remain the basic characteristic of the Fulani. Consequently, I suggest that the national government establish, at geographically convenient locations, transition shoots: that is, the construction of fenced corridors that link intensive farming areas with extensive grazing areas. Transition shoots are necessary because the disputes (and the reviving of past hostilities) that do occur arise over crop or cattle damage and the reduction of pastureland due to the expansion of farmland, mainly by recently converted herdsmen. Solving these problems shouid command top priority.

On the other hand, to emphasize the urgency of the land problem does not mean that the Hausa–Fulani relationship is degenerating, or is not working. On the contrary, it continues to be most successful, at least regarding survivability of nomads and farmers. When we consider moisture availability, soil fertility, and manpower requirements for both grazing and agriculture, it becomes clear that without cooperation in the social, economic, and political realms neither of these groups would exist at the standard of living they now enjoy. But they would not become extinct. In other words, I am suggesting that proto-cooperation, not symbiosis, best describes the Hausa–Fulani relationship. In fact, in certain aspects of their relationship, such as the economic aspect, one is tempted to draw an analogy between the Hausa–Fulani relationship and the lichens. In the case of the lichens, algae and fungi are so closely associated that botanists find it convenient to consider the association as a single species (Odum 1963, p. 108). I have suggested that under certain situations the Hausa and the Fulani are perceived as one ethnic group. But to consider this mutualism is misleading. Instead, I suggest, they are evolving towards mutualism.

Notes

1 No attempt is made here to exhaust the views on this point. It is important to notè that in the first half of the 19th century the Macina State of Sheku Ahmadu seemingly implemented a policy of sedentarization of the nomads (see Johnson 1976 and Vincent 1963) and those involved in transhumance were given state protection (see Gallais 1967).

2 Pasture regulation and allocation of cattle routes to accommodate Fulbe transhumance around Macina and the interior delta of the Niger is well known (see Gallais 1967 and Imperato 1972), although it is considered unique to the Macina area (see Swift 1977).

References

Allee, W. C., A. E. Emerson, O. Park, T. Park and K. Schmidt 1949. *Principles of animal ecology*. Philadelphia: W. B. Saunders.

Anthony, K. R. M. and B. F. Johnston 1968. *Economic, cultural and technical determinants of agricultural change in tropical Africa. Field study of agricultural change, northern Nigeria*. Prelim. Rep., no. 6. Food Research Institute, Stanford University.

Berry, B. J. L. 1964. Approaches to regional analysis: a synthesis. *Ann. Assoc. Am. Geogs* **54**, 2–11.

Buchanan, K. M. and J. C. Pugh 1955. *Land and people in Nigeria: the human geography of Nigeria and its environmental background*. London: University of London Press.

Gallais, J. 1967. *Le delta intérieur du Niger*. Davar: Institut Fondamental d'Afrique Noire.

Haswell, M. R. 1953. *Economics of agriculture in a savannah village*. London: Her Majesty's Stationery Office for the Colonial Office.

Hawley, A. H. 1968. Human ecology. *Int. Encycl. Social Scis* **4**, 328–37.

Hawley, A. H. 1971. *Urban society: an ecological approach*. New York: Ronald Press.

Hill, P. 1966. Landlords and brokers. In *Markets and marketing in West Africa*. Edinburgh: Centre of African Studies, University of Edinburgh.

Hodgkin, T. 1960. *Nigerian perspectives*. London: Oxford University Press.

Hogben, S. J. and A. H. M. Kirk-Green 1966. *The emirates of northern Nigeria*. London: Oxford University Press.

Hopen, C. E. 1958. *The pastoral Fulbe family in Gwandu*. London: Oxford University Press.

Imperato, P. J. 1972. Nomads of the Niger. *Nat. Hist.* **81**, 61–8, 78–9.

Johnson, M. 1976. The economic foundation of an Islamic theocracy – the case of Macina. *J. Afr. Hist.* **17**, 481–95.

Johnston, H. A. S. 1967. *The Fulani Empire of Sokoto*. Ibadan: Oxford University Press.

National Academy of Sciences 1965. *The science of geography*. Report of the *Ad Hoc* Committee on Geography. Nat. Res. Council Publ., no. 1277. Washington, DC: National Academy of Sciences.

Odum, E. P. 1963. *Ecology*. New York: Holt, Rinehart & Winston.

Scott, E. P. 1974. *Indigenous systems of exchange and decision-making among smallholders in rural Hausaland*. Michigan Geogr. Publ., no. 16. Ann Arbor: Department of Geography, University of Michigan.

Smith, M. G. 1960. *Government in Zazzau – 1800–1850*. London: Oxford University Press.

Smith, M. G. 1967. A Hausa kingdom: Maradi under Dan Baskore, 1854–75. In *West African kingdoms in the nineteenth century*, D. Forde and P. M. Kaberry (eds). London: Oxford University Press.

de St Croix, F. W. 1945. *The Fulani of northern Nigeria*. Lagos: Government Printer.

Stenning, D. J. 1959. *Savannah nomads*. London: Oxford University Press.

Stoddart, D. R. 1967. Organism and ecosystem as geographical models. In *Models in geography*, R. J. Chorley and P. Haggett (eds). London: Methuen.

Swift, J. 1977. Sahelian pastoralists: underdevelopment, desertification, and famine. *Ann. Rev. Anthrop.* **6**, 457–78.

Thompson, V. and R. Addoff 1958. *French West Africa*. London: George Allen & Unwin.

Turnbull, C. M. 1972. *The mountain people*. New York: Simon & Schuster.

Vincent, Y. 1963. Pasteurs, paysans, et pêcheurs du Guin-balla (Parti Centrale de l'Eng du Bara). In *Nomades et paysans d'Afrique noire occidentale*, P. Galloy *et al* (eds). Publ. Geogr. Inst. Fac. Letters Sci. Human. Univ. Nancy, no. 3.

4 *Nomads, farmers, and merchants: old strategies in a changing Sudan*

ANDREW W. SHEPHERD

Introduction and summary of the argument

In the Sudan the Savanna–Sahel drought of 1968–74 was less severe and its impact less destructive and tragic than elsewhere in the Sahel. Nevertheless, it will be argued here that it was an experience that turned the screw a few threads further into peasant and pastoral society, and accelerated the serious and widespread, though gradual, erosion of natural resources in the Sudan. The argument in this chapter will also be that the drought was an experience that old state-designed land-use strategies were ill equipped to cope with, but which has not led to any significant positive re-evaluation of these strategies, either by the state or by peasants. Land-use strategies in the early 1980s remain fundamentally the same as they were in the 1960s. Indeed, the state is locked into the same pattern of social relationships that characterized the period just prior to independence in 1956, and it interprets the recent drought and natural resource problems from that vantage point. Solutions offered by the state have been from a similar perspective, and have in any case gone practically unimplemented. The mass of farm peasants and pastoralists are unable to make the continuing investments required to halt drought-aggravating processes, either because of poverty, or because some investments are necessarily collective and would therefore depend on state intervention. The result is a persistence of the old state-designed land-use practices.

The state, then, has continued to pursue a highly extractive pattern of development, extracting revenue from the state-sponsored irrigated sector and allowing the extraction of food surpluses and capital by private capitalist farmers by development of the country's vast resources such as its deposits of clay soils. The agricultural peasantry and the pastoralists operate under pressure of land scarcity in many areas of the country. Extraction of value 'from the soil' can become their only option short of migration, or combined with migration, to maintain customary standards of living. Wealthier peasants may be able to maintain fertility-enhancing land-use strategies, because they experience less pressure of land scarcity, but this option is increasingly employed at the expense of poorer peasants, especially as wealthier peasants accumulate more and more land.

The 1968–74 drought undoubtedly contributed to the general concern for food security. The major function of the large irrigated schemes has been to produce revenue and foreign exchanges for the state, which is highly dependent financially on this sector. These schemes have in fact become significant producers of food crops for the local market, partly through planned diversification from cotton into wheat, groundnuts, and vegetables, but partly because tenants have seized every opportunity, legal and illegal, to grow staple foods and rear livestock. Paradoxically, it is partly the existence of the large-scale irrigated farm sector which insulates the Sudan from major fluctuations in staple grain production. This means that it will never suffer the acute famine conditions that other Sahelian countries have during prolonged drought periods. This is true despite the fact that the large-scale irrigated sector has deprived peasants and pastoralists of access to land and other natural resources, principally water. Partly it is also the dramatic development of large-scale, capitalistic, rainfed mechanized *dura* (sorghum) production on approximately 7 million acres of the plains of central Sudan which has insured the country against famine, and provided something of a stock of food to feed the hundreds of thousands of refugees who have poured into the Sudan during the past decade. This sector of production, discussed further below, has exerted an even more profound squeeze on peasant and pastoral economies. The paradox is completed by the greatly increased production and marketing of export (cash) crops and livestock during the last two decades in the rainfed, peasant sector, despite pressure on land and an absence of state investment in the small-scale farm sector. On the other hand, the state is now dependent on this peasant farming sector for up to 40% of its foreign exchange earnings. Nevertheless, the dangers of pursuing an extractive agricultural development policy in fragile ecological zones have been experienced, if only mildly, but the lessons seem not to have been digested or comprehended. The interests of national and international investment capital lie elsewhere and prevent sympathetic concern for the country's human and natural resources.

The drought

The severity and impact of the Savanna–Sahel drought of 1968–74 in the Sudan was far less than in other parts of West Africa. There is agreement (Government of the Sudan 1976), however, that the drought was a period when processes of environmental deterioration went into top gear, but that the fundamental causal processes, those related to human activities, were long in place and continued even after the drought. Similarly, the significance of the drought itself, *vis-à-vis* long-term underlying processes elsewhere in the Sahel, has also been questioned (van Apeldoorn 1981;

Garcia 1981). Nevertheless, it is important to pay some attention to the short-term causes and effects of the drought itself before going on to discuss the long-term causes and effects.

The period of the mid-1960s to mid-1970s has been one of low rainfall in the Sudan as elsewhere in the Sahel, forming a 'dry phase' in the long-term rainfall cycle, which followed a long 'wet phase' in the cycle (Ibrahim 1980, p. 15). During this dry phase, troughs in annual precipitation were experienced in different years, in different parts of the Sudan, and in different rainfall distribution which was somewhat localized. For example, there may be 200–400 km between the 200 mm isohyets in a wet and dry year (Ibrahim 1980, p. 13). The main area affected by this localized rainfall was western Sudan (Kordofan and Darfurl Provinces), which is the focus of this chapter. Administratively, these Provinces (now 'Regions') along with Kassala and the Blue and White Nile are all 'northern' provinces in the Savanna–Sahel zones.

The known short-term effects of the drought were several. Tree and vegetation mortality was undoubtedly high in many areas, and well water levels dropped (Digernes 1977, el-Faki 1977). Crop yields were extremely low during the periods 1971–2 and 1974–5 (Ministry of Agriculture 1977). Pastureland, especially near more northerly water points, became overgrazed and many animals died. The short-term effect in the Savanna–Sahel zones was sheer devastation.

The urban centers of western Sudan acted as hosts to the 'refugees' from the drought. Elsewhere, substantial new settlements of nomads, mainly those who had lost their animals, and of peasants, those who could no longer farm, grew up (Pons 1980). Some of these new settlements were temporary, as many 'refugees' subsequently attempted to return to their former land and life-styles. This option was much more difficult for livestock owners than for farmers, given the then high price for livestock. However, some 'refugees' stayed, became permanent wage-laborers, self-employed, or moved on to other towns. They did not return to their nomadic life-style.

Some large-scale voluntary resettlement also occurred. For example, some of Zaghawa livestock owners and millet growers of Northern Darfur moved south to the *qoz* (sandy) areas of Southern Darfur and became peasant farmers or wage-laborers (F. N. Ibrahim 1981, pp. 103–4). Other Zaghawa settled in Nyala and elsewhere in Southern Darfur where they became traders, apparently successfully competing with the established *jellaba* trading communities (Abu Sin 1980, p. 368). The Zaghawa voluntary resettlement was part of a much larger movement. One calculation claimed that during the drought years 30 or 40% of the population of Northern District of Northern Darfur (Dar Meidob) migrated southwards (Hales, undated work) where they established entirely new settlements in Southern Darfur.

The most widespread short-term reactions of peasant cultivators to an unusually dry period was to increase the area cultivated, at least where this was possible, by cultivating fallow land or pastureland. Cultivators with a large labor supply, whether this be family (women and children) or hired, are better able to withstand a short period of drought (F. N. Ibrahim 1981, p. 102) than those experiencing a labor shortage. The result of these practices is that during relatively short periods of unusual dryness, maximum areas of soil are exposed to wind and water erosion. The impact of wind and water erosion on their new farm plots is compounded by the methodical weeding of millet farms (Ibrahim 1980, pp. 2–3). The practice involves the removal of all competing grasses. This is clearly a rational response by the farmer who aims at ensuring subsistence for himself and his family. It is also commercially rational since the price of millet and other grains is high in drought years. Those families who are unable to expand their areas of cultivation to compensate for low yields are probably the ones who constitute the majority of migrants in dry years. The use of fallow land and pasture for cultivation does permit a degree of permanency during short, dry periods, but at the cost of environmental degradation.

Similarly, livestock owners with large or small transhumant herds are also better able to exploit distant pastures and water supplies than settled herders and can survive short periods of drought. Small herds are often composed of cattle and goats. They are less able to withstand drought and thirst than camels and sheep, which is why these are the main constituents of large herds in this semi-arid area (see Ch. 8).

During short-term drought periods, settled herds experience two kinds of difficulties. First, animal losses are especially high since they suffer from extreme overgrazing and slow vegetation regeneration, which usually occurs in proximity to permanent water supplies. Secondly, prices for the more drought-resistant sheep and camels are very high because former cattle and goat owners seek to buy them but owners are usually unwilling to sell (Hales, undated work). As a last resort, settled herders and other destitute peasants, both men and women, turn to wood and grass (fodder) cutting to replace lost income or subsistence (F. N. Ibrahim 1981, pp. 101–2). This practice, however, contributes to rapid environmental degradation.

Surprisingly, the impact of the drought, or indeed its basic features, have received comparatively little academic attention in the Sudan, relative to other Sahelian countries. This compares with Nigeria (van Apeldoorn 1981, p. 2), where the underlying processes contributing to desertification have received much more attention, reflecting the widespread opinion that drought is merely an aggravation of well established processes of deterioration in the balance between man and resources (this important point is discussed in Chs 1 & 6).

Underlying social and ecological processes

The immediate consequences of drought are reflected in longer term changes in society and ecology. Thus vegetation cover is known to have changed substantially during this century in the Savanna–Sahel zones. Tree cover has been markedly reduced (el-Faki 1977); certain palatable and perennial grass species have been removed from wide areas, and often completely from areas of livestock concentration, especially around permanent water supplies (Harrison 1955, el-Hassan 1981, el-Tayeb 1981). Crop yields show secular downward trend, whereas cultivated areas, conversely, evidence upward trends (Government of the Sudan 1976). In essence, the natural landscape is being replaced by a man-made one.

Urbanization has continued during this century at a steady pace (el-Arifi 1980) with an increase in the rate since independence, due mainly to the expanding gap in standards of living between town and countryside, and to the expanding rate of capital accumulation by the ruling urban merchant class. Fuel, fodder, and food demands of urban populations have all put increased stress on the mainly untransformed methods of forest, livestock, and agricultural production. Desertification, due primarily to woodcutting to meet fuel demands in large settlements of the Savanna–Sahel zones, is the most observable, dramatic effect (Digernes 1977, F. N. Ibrahim 1981) of rapid urbanization in the Sudan (about 25% of the total population lived in urban centers in 1982).

In rural society, it is the *concentration* of population and human activities, as well as their absolute growth, which is widely seen as the major cause of desertification and general ecological stress. Argument continues as to whether settled cultivators or transhumant livestock owners are more to blame. The fact is that the two 'sectors' form a social and economic continuum (Hunting Technical Services 1974, Annex 5), with continual movement between them, and a complex of relationships of symbiosis and conflict (also see Ch. 3). Thus sedentary cultivators invest a portion of their profits in livestock (Haaland 1977); and pastoralists also cultivate. Indeed, poorer pastoralists, or those who lose their herd viability, often depend heavily on cultivation. Where there are investment opportunities in agriculture, as on the large-scale irrigation schemes and in mechanized, rainfed farming, pastoralists have readily taken these up. To answer the question 'Why have farm peasants and pastoralists taken part in the depletion of their own resource base?' requires a historical analysis which is national in scope. This is what the remainder of this chapter attempts to do.

Transformation of the economy

During this century, the penetration of the 'market economy' and the creation of a dual economy have been the principal external economic forces influencing land-use patterns in the peasant economy of the Sudan. The establishment of the Gezira Scheme in 1925, for example, remains the central fact of modern Sudanese economic history, but one which has done so in a highly uneven way, by drawing labor from the poorest and least commercialized areas, from the northern provinces and from West Africa generally to serve as temporary and permanent proletarians in the central, modern economy (Duffield 1979, O'Brien 1980). The mechanisms used by the colonial regime were the classical ones: taxation, marketing of consumption goods, prohibition of local cotton cultivation, persuasion and forced recruitment, and even the restriction of development outside the Gezira Scheme (O'Brien 1980, ch. 8). This latter effort to transform the peasantry, combined with the attempt to conserve or recreate peasant modes of production outside the irrigated sector after the British reconquest, and the encouragement of settlement by West African and Chadian immigrants, has served to keep farm wage levels low. Recent intensified mechanization of production in the state-sponsored irrigation schemes can be interpreted as the most effective method of planned wage reduction, necessary only from the perspective of resolving the revenue crisis without extensive damage to the state.

The impact of wage-labor migration on the national economy has been variable. It has offered significant and widespread cash-earning opportunities to farm peasants and their families, often complementing their home productive activities in agriculture. This opportunity is quite convenient for farm peasants since the main operation for which wage-labor is required in the modern sector is cotton picking, which occurs late in the growing season when their own work is done. On the other hand, the absence of male labor has sometimes had especially adverse effects – fatherless families, the individualization of production and consumption, women forced into beer brewing and prostitution when remittances fail to come in (el-Faki 1977), a decline in the gum arabic economy where the prices offered for gum tapped do not equal migration wages, and consequent decline in soil fertility as gum trees are cut down (el-Din, undated work) – but this would apply more to long-term migration than to short-term or seasonal migration. Seasonal migration is quite beneficial to farm peasants because it has tended to supply cash to their families to purchase agricultural inputs and to hire labor, as well as goods for consumption. However, when seasonal migration fails to complement home production in time, and cash needs nevertheless force seasonal migration, home production suffers (al-Bashir 1980) and famine may result.

This argument masks the main significance of migration, which is, first, the partial conversion of an independent peasantry, mainly those in western and central Sudan, into an agricultural semi-proletariat; and secondly, that it is only certain segments of the peasantry which have by and large been so converted: the proletariat is overwhelmingly composed of western

Sudanese and *Fellata* (originating in West Africa), although sundry other marginalized groups also participate, for example, the Beja and Beni Amer in eastern Sudan. Class formation has thus taken on a strong regional dimension in the economy of the Sudan.

The operation of the capitalist market has been vastly extended into the countryside, not only in consumption goods, originally cloth, tea, and coffee, but now also radios, crockery, and other consumer items. With the building of a basic infrastructure – principally the railway to El Obeid and then Nyala and Wau and its associated feeder roads – and the gradual spread and consolidation of trading or *jellaba* communities, tracing their origins to the Northern Province and the Gezira, eventually settling throughout northern and even southern Sudan, trade in agricultural products expanded well outside the riverine areas. Gum arabic, livestock, and cotton, principally in the Nuba Mountains, were the main commercial products developed during the colonial period. Groundnuts, sesame, sorghum, *kerkadeh*, tobacco, and vegetables are now widely grown on a commercial basis and for export. The revenues they generate are now crucial to the wellbeing of the government. The President of the Sudan recently publicly admitted that as much as 40% of export revenues are derived from these crops that are produced in the 'traditional sector', and up to 20% of Gross Domestic Product is produced by livestock rearers alone. Despite the extensive expansion of the modern sector during and since the colonial period, the state remains embarrassingly dependent on the peasant sector, which grew substantially during this time. To a very great extent, this peasant expansion has happened in response to market stimuli and to the operation of the *jellaba* diaspora. The state, on the other hand, has played a minor role beyond the provision of basic services, the imposition of government structures to maintain national cohesion and security, and the protection of the local pre-eminence of various *jellaba* groups.

The results of enhanced commercialization of the countryside are many, complex, and variable. For example, arable land became scarce, the fallow period was reduced, the monetization of land increased, local labor markets and a landless labor pool developed, new patterns of labor migration developed (e.g. Bahr el Ghazal to Southern Darfur and Southern Kordofan), and the peasantry became differentiated. Because of the oligopolistic structure of trade (Abakr & Pool 1980), the major beneficiaries, in terms of capital accumulation, have been the merchants, especially those with wholesaling capacity and links to Khartoum. In testimony to their commercial strength, the major towns of western Sudan are now mainly trading entrepôts, dominated economically by *jellaba* communities. Although certain inroads have been made into the established structure of trade in western Sudan by Western groups, the position of the *jellaba* remains strong, although strengthening regionalistic and anticapitalist sentiments have placed them in an increasingly precarious situation in the last 2–3 years.

Part of the expansion in crops grown for export can be explained by the long-term decline in the producers' internal terms of trade. Internationally, commodity prices have declined *vis-à-vis* industrial goods, and this has been reflected in the local market. This development, coupled with *sheil* (crop mortgage) arrangements entered into by many producers and the low farm-gate prices offered in any case by oligopolies of merchants, have meant very small increases in producer prices over the years. Logically, producers have had to 'run to stand still'. Few have been able to accumulate much capital. Where they have, it tends to go into livestock investment for reasons of security and capital growth (Haaland 1977, p. 180) in a time-honored and still valid pattern, despite the degradation of pastures. Only in Kassala and Blue Nile Provinces have second-hand tractors been an alternative investment for peasants.

In those agricultural communities settled by Nubians in western Sudan (e.g. the Southern Kheiran villages; Manger 1981), they tend to do better from agricultural changes than the local people. Whether this is due to entrepreneurial talent (Hunting Technical Services 1974, Haaland 1980, p. 66), or to superior access to the state machinery and historical factors, is a controversial issue where the truth lies probably in a combination of all of the above. Merchants are now increasingly investing directly in agricultural production in Western Sudan, often acquiring the best quality and best located land (Abakr & Pool 1980, pp. 194–6). The gradual conversion of the peasantry to export crop production and the growing merchant investments in agriculture are transforming the whole Sudan into an extractive agricultural economy.

Commercialization of both land and crops, coupled with population growth, both human and animal, has created pressure on cultivable land and pastureland on an unprecedented scale. Areas around water supply points are heavily overgrazed. The situation is acute in the northern locations. Studies show that in Southern Darfur in Dar Rizeigat where livestock concentration is high, the livestock carried by pastureland is almost 10 times that required for sustained production (Khogali 1979, p. 57) in the northern areas. In Dar Meidob (Northern Darfur) stocking rates during the 1970s probably matched carrying capacity; yet there was local overgrazing, especially around settlements and water points (Hales, undated work). The effect is that forest cover is removed and sand dunes spread southward (Digernes 1977). Elsewhere, farm fallow periods are reduced and gum arabic ceases to be part of the rotation (Digernes 1978, el-Din, undated work), thereby losing its beneficial nitrogen-fixing and soil-stabilizing qualities. Most seriously, in the short term, shortages of water, competition over grazing land and conflicts between herders and cultivators have led to widespread disorder in rural society. Fatal incidents are common; almost annually intertribal wars flare up in which hundreds may be killed (Abakr & Pool 1980). Rural administrations are no longer capable of preventing such occurrences (Karam 1981). The extent of the crisis is not yet clear; nor are solutions. What is clear is that the frequency and intensity of these conflicts are associated with resource shortages that are rooted in long-term social and ecological changes in the Sudan.

At the same time, the state is increasingly dependent on peasant agriculture and livestock production: the pressure to extract more directly from the peasant economy will increase as budget and balance of payments deficits rise during the 1980s. In theory this should lead to greater state investment in rural development, especially in Northern Darfur. Indeed, there are signs of increased governmental interest during the 1970s. This interest was expressed mainly by the large number of government-sponsored rural development aid projects, all of which commit some national resources. These investments are trifling, however, when viewed against the major public undertakings in the modern agricultural sector (Shepherd 1982a, 1983).

On top of these long-term, stress-producing processes, large-scale mechanized *dura* (sorghum) production is expanding in Western Sudan (O'Brien 1977, al-Bashir 1980, Saeed 1980). If the expansion is substantial enough, there is no reason to believe that its impact will be different from that experienced in central and eastern Sudan.

State capitalism: the irrigated sector

The extensive government-sponsored irrigation farming sector, now covering about 3.7 million acres, has its roots, as we have seen, in the colonial political economy, and specifically in the 'success' of the Gezira Scheme (Gaitskell 1959; Barnett 1977, 1981; Salam 1979). What has typified the postcolonial era has been the dramatic expansion of this state-sponsored sector on more or less unchanged principles. Centrally run, these irrigated schemes are divided into tenancies allocated by the irrigation authority which closely controls what is produced, how and when it is produced, and who purchases the crops, such as cotton for export. Although some adaptations have been made to the Gezira Scheme model to suit local circumstances elsewhere in the country (e.g. the cultivation of wheat in New Halfa by the Halfawiyeen tenants and the current attempts to develop livestock tenancies in the Rahad Scheme), the above fundamentals, which are designed to ensure a dependable source of state revenue, have been adhered to. Irrigated agriculture in all its basic features has a sameness to it, due to the widespread adoption of the Gezira Scheme as a model, that may be unique to the Sudan.

Far from reducing reliance on small peasant tenants, as some would wish (e.g. International Bank for Reconstruction and Development 1966, p. 71), recent developments in irrigated farming such as the Rahad Scheme, have confirmed the pattern (O'Brien 1980, pp. 186–7). Partly due to administrative inertia, the persistence of the pattern may also be due to the fact that government revenues can only be assured by this pattern. Indeed, the interest of the state is served by maintaining the status quo. For instance, capitalist farmers (the obvious alternative, sometimes advocated by aid agencies) would be much more difficult to control and extract revenues from, given the much greater resources, both material and political, at their command. The promotion of capitalist farming has thus been confined to rainfed food production in which the interests of the state are more indirect, where a concern for food production, for national security and political order, and for the class interests of families and social groups who dominate the state apparatus, rather than a concern for revenues and foreign exchange, is paramount. For the latter, the model devised by the British for cash-crop production remains more attractive, even though the original British assumption that the tenant would be a cultivating peasant of some prosperity has not worked out in practice. In fact, tenants of irrigated farm plots have historically acted like foremen, organizing hired labor to work their tenancies. They have been part of a hierarchical, bureaucratic structure which would not suit the mode of operation of capitalist farmers.

There is a contradiction in the pattern. The extraction of revenue has resulted in the majority of tenants remaining poor. Many are constantly in debt (Barnett 1977, Pearson 1980). Poor tenants or subtenants (O'Brien 1980, p. 198) are unable to invest adequate working capital in their tenancies, so yields remain low. Reduced crop yields by this relatively large group reduces revenues from the entire sector, and has been a major factor contributing to Sudan's large and chronic trade deficit. With the country's debt burden mounting and the continued expansion of government into the farm sector and the economy generally, the pressure for maximum revenue extraction remains immense. Yet to increase the overall revenue base, the rate of extraction from direct producers must be reduced. In the past it was possible to expand the revenue base laterally, by irrigating new land. The New Halfa Scheme and the Managil extension of the Gezira did indeed generate new revenue initially, but now none are consistent revenue earners. New strategies must be employed.

The failure to produce consistent and sufficient revenues even from an increasingly irrigated agriculture has led the Nimeiri regime to an extraordinary degree of dependence on Western and Arab loan capital for the physical, and to a lesser extent, socio-economic rehabilitation of the state sector. The interests of international capital and the Sudanese elite, as presently constituted, are mutual insofar as the goal of rehabilitation is to maintain the external, export orientation of the irrigated sector. The interests of the tenants, on the other hand, and of the majority of the Sudanese population, lie in maximizing food security (Barnett 1977, pp.

101–15). They act in significant ways in their own interests. Irrigation water, when scarce, for example, is deviated from cotton to food production by the tenants, thereby reducing production of revenue-generating crops in the irrigated sector of the economy.

The interests of nomads also come sharply into conflict with the export orientation of the state irrigated sector. Irrigation schemes lie alongside rivers, sources of dry season watering for herders, and they occupy former dry season browse and grazing lands. Nomads have tried to adjust. In eastern and central Sudan they now rely on irrigation canals for drinking water, much to the irritation of the Irrigation Department; they now also rely on crop residues, which any animal may graze legally after March 15, and on green or ripening crops, which are grazed illegally before March 15, much to the irritation of the scheme management.

Food security for livestock in central and eastern Sudan has, since the drought, increasingly been assured also by the mechanization of cultivation amongst the nomads themselves, and the widespread practice of buying animal fodder concentrates during the dry season (Shepherd 1982b). However, the integration of irrigated crop production and the livestock economy remains partial, unplanned, and often illegal, a situation that does not meet the interests of the large nomadic population of central Sudan. Since many of the nomads are also tenants on the irrigation schemes (Sorbo 1977, O'Brien 1980, Pearson 1980) who suffer from the extractive policies described above and from the constraints on their food security, the interests of the state and of the masses remain widely divergent and conflicting.

*Capitalist, rainfed mechanized sorghum (*dura*) farming* Thus far, we have discussed developments associated with the state-sponsored irrigated farming schemes where crops are produced primarily for export and the generation of government revenue. Government intervention into the rainfed agricultural sector is well established but fraught with controversy, mainly as it affects the rural society.

From small beginnings in the 1940s and 1950s, state-subsidized, mechanized rainfed agriculture on the clay silt plains of central and western Sudan has expanded to cover about 7–8 million acres (Fig. 4.1). Government-demarcated, mechanized farms cover about 5 million acres and an additional 3 million acres have been cultivated without permission of the government and are undemarcated (Bryant 1977, p. 43). Sorghum (*dura*) alone is produced on most of these farms, which in turn is sold in the cities and towns. The bulk of mechanized farming is carried out on privately owned and operated farms, although there are some state farms, a few state-run cooperatives, and some use of tractors by peasant farmers. The majority of farms are capitalist in organization: an entrepreneur (usually a merchant) brings capital and labor together and manages his farm for

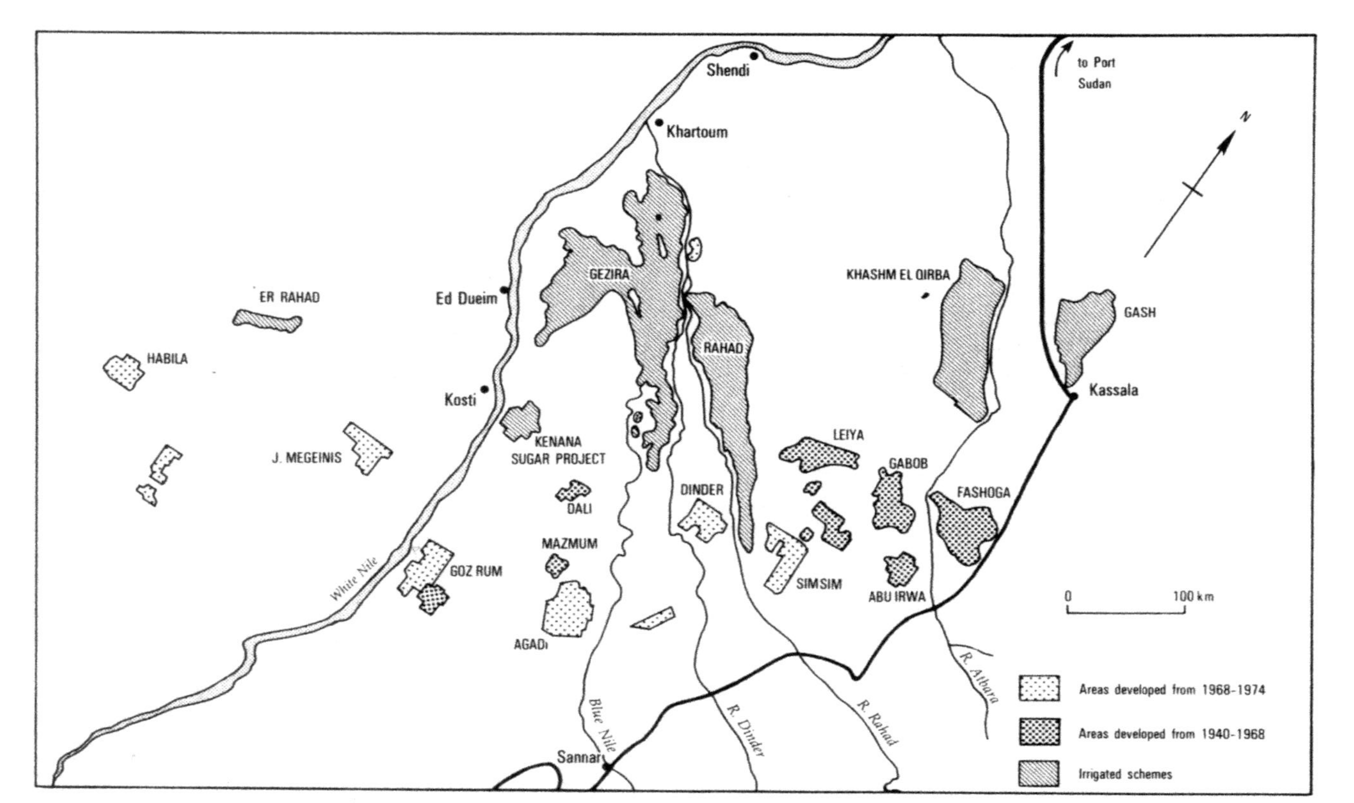

Figure 4.1 The expansion of demarcated mechanized crop production near Gedaref. (*Source*: Mechanized Farming Corporation, Gedaref 1982.)

private profit. Profits are extracted by the merchants from the exploitation of local peasant and Western African or refugee immigrant labor from the land for which exiguous rents are paid to the state, and from its former users who have been ruthlessly expropriated by the state with little compensation, a process of primitive capital accumulation, the main beneficiary of which is the merchant or *jellaba* class (Simpson & Simpson 1978; O'Brien 1980, ch. 5; Shepherd 1983).

During the 1950s and 1960s expansion of large-scale, state-subsidized mechanized farming was slow. Even then, however, there was some awareness in government circles of the strain it placed on the peasant economy, and of the need to incorporate the peasantry as beneficiaries of development – something which was never done (Ministry of Agriculture, undated report).

In 1968, the Mechanized Farming Corporation (MFC) was established to supervise and control mechanized farming. The purpose was to secure a World Bank loan of $16.25 million to shore up the rainfed farming sector. This loan, and the Bank's second loan in 1972 of $11.2 million, went to the so-called 'supervised sector' where the MFC provides credit for machinery and is supposed to supervise production on 760 000 acres of rainfed land. In addition, the MFC was to supervise the 'private sector', where farmers raise their own capital, often from the state Agricultural Bank of the Sudan (ABS).

By 1967–8 the demarcated, rainfed cultivated land amounted to 1 993 000 acres, mostly in Kassala and Blue Nile Provinces (Simpson & Simpson 1978, p. 105). Under the MFC, the demarcated land put into cultivation has increased to approximately 5 million acres, with large expansions in Southern Kordofan, White Nile and Upper Nile Provinces as well as in Kassala and Blue Nile.

The establishment of the MFC has, in fact, laid the services of the state at the feet of the lessees. Leases were lengthened to 25 years and farm sizes were increased to 1500 acres to incorporate a 500 acre fallow. The MFC acted as the channel for World Bank credits to farmers to cover the costs of mechanization and land clearance, which were to be repaid at well below market rates of interest. The MFC also offered mechanized land clearance facilities, the cost of which was to be repaid over a generous 25 years, and in addition it provided centralized machinery maintenance facilities.

Outside the World Bank schemes (at Gedaref (Simsim, Um Seinat) and Habila), the ABS has given farmers a comparable level of financial assistance: up to 70% of total production costs to be repaid at 9% interest over 5 years in the case of fixed assets and land clearance, and over 18 months in the case of seasonal costs. There has been a very low capital repayment performance in both the cases of the MFC and ABS (Due 1980, p. 229), leading to losses experienced by both organizations running to millions of Sudanese pounds. Partly as a result of this government support, mechanized farming became extremely profitable and demand for leases was excessive.

Returns to capital in mechanized farming are difficult to assess or to generalize about, but they have usually been found to be high (O'Brien 1980, pp. 110 ff.; Shepherd 1983).

As far as the peasantry are concerned, some relief has been provided by the expansion of the national labor market (O'Brien 1980, chs 12–13). Indeed, it is still through the labor market that 'modern' and peasant sectors of the agricultural economy are integrated. But the effects on independent peasant production and survival are clearly negative. Figure 4.2 illustrates the vast chunks of land that have been removed from the peasant economy, a development that will mean greater numbers of peasants moving into the national labor market.

More obvious results of this expropriation are: wholesale loss of forest (which is also browse), including government forest reserves, a small compensatory factor being that charcoal is as a result cheaper than it would otherwise be for urban consumers; loss of grazing, though this is to some extent mitigated by the new reliance of livestock on crop residues, and

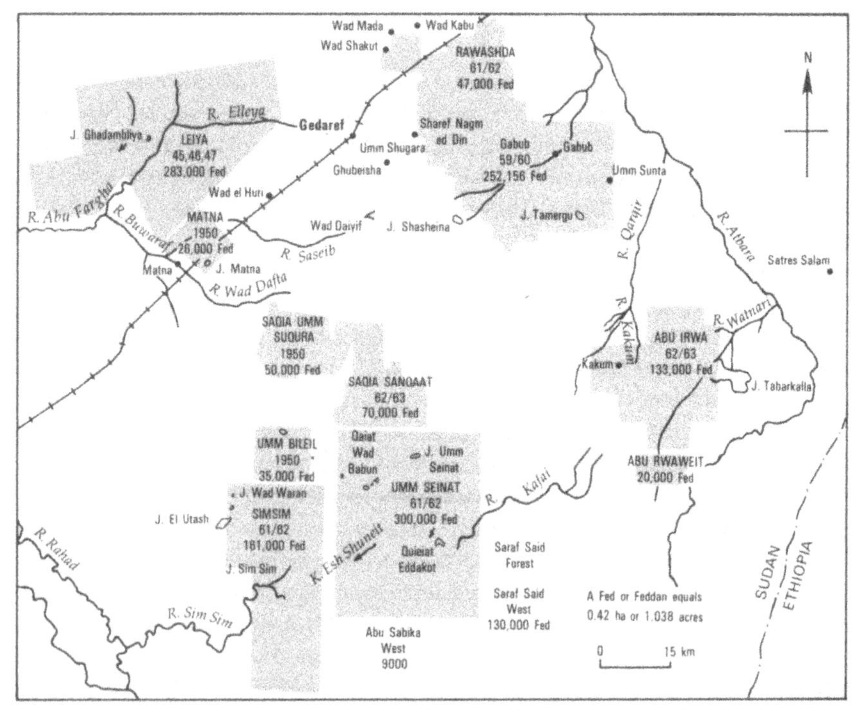

Figure 4.2 Major irrigated schemes and areas of demarcated mechanized farming in East Central Sudan.

illegal grazing and watering on irrigation schemes; the loss of room for bush-fallowing, leading to lowered yields and massive weed problems in peasant areas; and a substantial degree of differentiation in local peasant communities (Shepherd 1983).

Clearly, there has been little adjustment by government of the major agricultural policies as a result of the 1968–74 drought. The potential for greater peasant, and even nomad, formal involvement in mechanized rainfed agriculture is clearly demonstrated by their willingness to hire tractors and, where possible, purchase second-hand tractors for use in cultivation. To ensure their participation, the government must be willing to organize the peasantry and above all be politically committed to safeguarding their general welfare. The government and the country's ranking groups have not demonstrated an inclination to do this. Similarly, the potential for balancing food, livestock, and export crop production in the state irrigated sector is also there. In the long term, a better balance in these three areas of production is inevitable, but in the short term it has been resisted by the vested interests in the present structure of production and distribution of the product. The burgeoning and intertwined merchant and bureaucratic classes are the main beneficiaries of both policies. The peasantry continues to suffer.

Government responses to drought

Water supplies The major practical form that investment in rural development in the Sudan has taken is the construction of facilities to supply water. Toward the end of the colonial period, a double concern for soil erosion and conservation on the one hand (Ministry of Agriculture 1944), and increasing marketed crop surpluses from peasant agriculture pending the development of viable capitalist farming methods (Jefferson 1955) on the other hand, led to an emphasis in investment in rural water supplies. This has been at the heart of most subsequent rural development work. Investment in water supplies in new, peripheral areas was to attract people and livestock away from the overcrowded, settled areas (the concern for soil conservation in general was in practice boiled down to this policy). The precondition for the development of more productive and settled farming, often on virgin land, was the provision of water. Water supply installations also proved to be a means of purchasing political support in the struggle leading up to Sudan's independence from the British, particularly from the traditional and conservative Ansar (followers of the Mahdi) based in Western Sudan. The centrality of water to both development and politics is an indication of its scarcity and importance in the Savanna–Sahel zones (Shepherd & el-Neima 1982).

Easy access to clean water being the item most urgently lacking for the majority of Sudanese at independence, politics since then has been partly an expression of competition for water supplies. The technical and developmental aspects of water supply investment have tended to be secondary. National political reputations have been built on water, and local political office is normally gained partly through promises of water. Such political impetus has undoubtedly resulted in the satisfaction of a substantial proportion of the original need.

During the 1950s land-use planning considerations were linked to the provision of water supplies in a more formal way than was the case earlier. However, these considerations were, and remain, site specific. In other words, only the immediate hinterland of a site was considered by the Land-use Officer when a request came for improvement in water supply. The officer's survey took into consideration the carrying capacity of the land, the vegetation cover, the socio-economic characteristics of the people, and their productive potential. But the considerations did not extend to the interaction between one site and another, which would have enabled area-based (i.e. regional) planning. The decision-making process was, and is, dominated by expressed needs, and political pressure. From an economic point of view, water is a prerequisite of commercialization in this semi-arid, low rainfall Savanna–Sahel area. Continued commercialization has, of course, served the interests of those who dominate the national and local economy. Concentration of population is the result of investment in water supplies, and sometimes other services. It is also a major mechanism of commercialization, allowing the formation of a producer market, adequate to the interests of local merchants.

Professionally, engineers and hydrogeologists have dominated agriculturalists and land-use planners who are associated with the provision of water supplies. Organizationally, the disagreements over policy as well as land for distribution and control became so severe as a result of the rapid development of the late 1960s and early 1970s that in 1975 rural development and land-use planning (the Soil Conservation, Land-use and Rural Water Programming administration) were finally separated administratively from the engineering and geological aspects of water supplies, although the Rural Development staff retained their planning function in relation to water supplies, and both remained in the Natural Resources Division of the Ministry of Agriculture. Since then there has been a much lower rate of investment in water, and rural development has sunk back to a low-level priority according to the 1977–83 Six-year National Plan, its staff and morale severely depleted. Today, the central organization has become a clearing house for aid programmes in rural development.

The Rural Development staff perceived only too clearly the adverse ecological effects of many water investments, whatever the social, political, and economic benefits (el-Hassan 1981, el-Tayeb 1981). The provision of large-scale and often maldistributed water supplies, it is alleged, has been a central element in creating the resource crisis in the peasant economy described above. This was one example of a by now familiar critique of unimodal development interventions, and Rural Development staff in the Sudan were not slow to take on board the ideology of Integrated Rural Development (IRD). However, the bureaucratic empire-building, which IRD was taken to imply by other departments, was seen as a threat and this has helped to undermine efforts during the 1970s to integrate the variety of rural development work carried out mainly in the Natural Resources Division of MAFNR (Ministry of Agriculture, Food and Natural Resources). Other undermining factors have been the shortage of funds allocated to the division, such that line departments have not been able to carry out existing functions, let alone take on new ones. The operation of aid agencies has also limited opportunities for integrating rural development activities.

Desert Encroachment Control and Rehabilitation Program This last undermining factor can be seen in the history of the government's Desert Encroachment Control and Rehabilitation Program (DECARP: Government of the Sudan 1976). Launched as the government's response to long-term processes of desert-creep, an attempt was made to get collective donor agency funding for the whole program. However, this effort broke down, and projects and subprojects of the overall program have had to compete for separate financing. Only a very few of these projects have as yet been funded and fewer still, if any, have been implemented. This reflects the lack of real governmental interest and commitment to nonprofit-making projects, however essential (Meillassoux 1974, p. 32), as well as the destructive effect of interagency arguments and jealousies (Sheets & Morris 1976). All these factors contribute to the 'administrative trap', in which the Natural Resources administration is carved up amongst several professions, making it quite incapable of dealing with interdisciplinary problems (Baker 1976, p. 248).

In fact, the programs were loosely integrated, both intellectually and administratively, and therefore arguably unsuited for joint funding. Projects put forward for funding were more or less a shopping list and were mostly extensions of existing unintegrated departmental work. In this respect, and even in the content of some of the projects, there is a strong resemblance to the 1944 *Report of the Soil Conservation Committee* (Ministry of Agriculture 1944). That committee's recommendations put forward, quite self-consciously, the pet ideas of the committee members. DECARP put forward its pet ideas of field department, which in one way or another trace their origins back to the 1944 report.

DECARP relied on the People's Local Government System (1971–81) to integrate, plan, and monitor its activities. However, this political and administrative system had already proved relatively incapable of environmental work (Glentworth *et al*. 1976, el-Arifi 1979, p. 39), although some officers (and many writers) believe that rural development work can be more easily coordinated at provincial or regional level than at central level. Basically, local governments in the Sudan do not have the funds to invest; their development budgets have depended on the central government (Ibrahim 1982), which has had its own budgetary difficulties during the 1970s, and has been unwilling to decentralize much investment finance.

Beyond this, DECARP evinced considerable idealism over questions of implementation. For example, the Gereih el Sarha Project was put forward as a successfully managed grazing scheme to be emulated by others (Government of the Sudan 1976, p. 96). A technical success it undoubtedly is, carrying capacity of pastureland having increased from 386 animal units per year in 1972 to 1390 in 1979 (author's fieldwork in North Kordofan, 1981). But it could not resolve the wider issues of nomad settlement (Thimm 1979, pp. 3–37), which remain unresolved.

In another instance, the development of town perimeters (Government of the Sudan 1976, pp. 100 & 211–12) were put forward without mention of earlier difficulties in establishing similar grazing management schemes at Babanusa (Adams 1982, p. 277). And, as far as water supplies are concerned, it failed to recognize their role in increasing pressure on grazing near the water supply as well as their role in increasing the overall numbers of livestock in the Savanna–Sahel zones. It is also recognized that improved veterinary provision (Government of the Sudan 1976, p. 98) has resulted in vastly increased livestock in this region. Better planning is obviously called for (Government of the Sudan 1976, 40), and it must include the integration of range conservation and management. The latter has seemingly never materialized, except for a short while in the mid-1970s when the National Resources Division of MAFNR and the Range Management and Forestry Officer were supposed to participate in the planning of water supplies.

The planning of water supplies in relation to expanding agriculture which was recognized to contribute significantly to arresting desertification (Government of the Sudan 1976, pp. 93–4), and in some areas its provision is probably the major cause (Digernes 1978, Ibrahim 1980), but curiously water supply is given little emphasis. There are no agriculturalists among the land-use planners. This omission probably reflects the degree to which agriculturalists and others (irrigation engineers, animal husbandry experts, and veterinarians) concerned chiefly with production, whose offices absorb the lion's share of MAFNR budgets, are insulated from Natural Resources professions and concerns. This derives from the splendid isolation with which development projects in rural Sudan have typically been designed and managed. 'Hinterland' problems are seen as the province of others – in this case, the Natural Resources Department.

The many good intentions that characterized DECARP have affected rural development aid projects generally in the Sudan (Adams & Howell 1979, pp. 514–18), including those specifically aimed at reducing ecological stress (e.g. Hunting Technical Services 1974, 1976). The ambitious Southern Darfur Development Project has thus, after many years of surveying and planning activity, crystallized into a small water supply and settlement program (a. J. Abdulla, personal communication). This is due to decision and financial delays by the aid agencies and the central government, to administrative practices in the country, and to the 'implementation bottlenecks' that aid agencies are quick to spot but slow to analyze as being the result of established political and administrative practices, including, of course, widespread corruption of regulations.

'Modernization' schemes Beyond investment in water supplies and other services (health, education, and veterinary health), a colonial interest in promoting small settlement schemes based on improved agricultural practices and served by infrastructural investments (Jefferson 1955, para. 63) has gradually been abandoned for a single-minded adoption of mechanization (Adams & Howell 1979, p. 511) in an attempt to impose change in the manner which has, to some extent, met the purposes of the state in Eastern and Central Sudan. Mechanization, however, may be dangerous for local farmers in remote areas (el-Tayeb 1981), mainly because of its unreliability, and in ecologically fragile areas, unless husbandry standards are very high (Adams & Howell 1979, p. 512). It requires attention to support facilities, local infrastructure development, and the land-use system, as discovered in the Nuba Mountains (Thimm 1979, p. 32). These have frequently been inadequate, leaving the peasants with the worst of both worlds. As in other aspects of rural development work pre- and postdrought policies show a high degree of continuity.

With few exceptions such 'modernization' schemes have had little impact. The principle exception – the Nuba Mountains Agricultural Production Corporation – is also the best organized: reflecting perhaps its concentration on cotton (dating back to the colonial period), and therefore potential revenue and foreign exchange extraction. For this even subsidies have been forthcoming (Thimm 1979, 32). Other schemes have been run by weak and marginal Natural Resources or cooperation departments as one of several; they have suffered from inadequate attention and resources, and disciplinary single mindedness.

Conclusion and comments

Food security A degree of national food security has in fact been achieved by the Sudan's present development strategy. However, private mercantile control over the distribution of staple food grains has meant that food deficit areas, such as the semi-arid zones, and parts of Southern Sudan, often do not receive the amounts of grain needed to maintain a reasonable stability of prices. Merchants will sell in areas where there is purchasing power. This means that food deficit areas tend also to be low income areas. The real tragedy is that low income areas are poorly supplied in drought years when livestock mortality is high and savings cannot easily be mobilized. Famine may result, not because of a scarcity of food, but because of poor distribution of available food.

In part, this is a problem of weakly developed communication infrastructure, leaving many regions relatively isolated from the national market. In part, it is also the result of a weak public sector intervention in the staple food grains market in relation to requirements. The consequence is that in dry years high grain prices create incentives to expand cultivated areas in the semi-arid zone. This is in addition to the logic of family food subsistence which pushes cultivation in the same direction. Greater public sector intervention would clearly be against the interests of the dominant merchant groups. There are also vital issues concerning production relations and the organization of production in both the rainfed mechanized sector and the state irrigated sector, which have an impact on national food security. But the question of food distribution, not scarcity, is probably the most important for the semi-arid zone.

Just as it is difficult to envisage greater and more effective public intervention in staple food grain distribution under the present political circumstances, it is also difficult to believe that production relations in the state irrigated and mechanized rainfed sectors can at present be substantially transformed in the interests of national food security. There is, however, a creeping reform of production relations in the state irrigated sector, induced by peasant and nomadic (or ex-nomadic) tenants who allocate time and water to food crops and animal fodder before they attend to their export cash crops. Tenants' unions also have a long history of bargaining for greater formal incorporation of staple food crops and animal fodder into rotations. These processes could be picked up and used by future governments interested less in revenue and more in the general public interest. Furthermore, the present activities undertaken by livestock keepers to integrate production of livestock with irrigated crop production could be experimentally legalized and formally incorporated in planned scheme operations. Considerable imaginative work is undoubtedly required to achieve such balances; experimentation is unavoidable.

Toward ecological balance? In the ecologically fragile zones, themselves outside the main investment areas of Central and Eastern Sudan, there are perhaps three ways in which the analysis of this chapter may contain lessons for future land use. First, the need for reafforestation is paramount. This is presently conceived uniquely as restocking the gum arabic belt, which stretches across the low rainfall Savanna–Sahel zones. The potential for revenue generation is what attracts the meager national and international finance available for forestry to the gum arabic (*hashab*) tree. Although such reafforestation would clearly be beneficial, there is the price structure to contend with. Low producer prices coupled with poor infrastructure and inadequate water supplies during harvesting, as well as poor communications, have all underlain the abandonment of gum arabic gardens in much of the Western Sudanese gum arabic belt. Furthermore, in many areas today, land scarcity and shortened crop rotations in the vicinity of dense settlements mean that holdings are much smaller than they were in the heyday of the gum arabic trade. Without changes in the structure and pattern of land holding and/or infrastructural investments, especially in water supplies, to make new lands accessible, it may be difficult to reincorporate gum arabic trees into the crop rotations. Other trees and forest products have also been neglected.

Secondly, there is a persuasive argument in favor of transhumance and livestock production over permanent settlements and cultivation in the Savanna–Sahel zones (Ibrahim 1980), especially to the north of the restocked gum arabic belt. The main reason for this is that only transhumance can adapt to variability of rainfall (also see Ch. 8). Also, transhumance has in the past been combined with a low concentration of population. Whether the social conditions for a return to transhumance exist is a question that requires detailed analysis, and has implications for development strategies in the higher rainfall Savanna–Sahel zones where any 'excess' population might need to be concentrated.

Thirdly, and related to both the above points, is a commonly perceived critique of sedentary animal husbandry, which tends to be more destructive of natural resources than transhumance. There is still a strong trend in West Sudan towards spontaneous settlement by livestock herders, for powerful social and economic reasons (see Chs 1 & 3 for more detail). It is not clear how the separation between livestock rearing and cultivation envisaged, for example, by plans for Southern Darfur could be achieved in practice unless the security and profitability of cultivation are greatly improved *vis-à-vis* livestock. Although separation of animal husbandry and cultivation may be feasible and desirable in some areas, clearly in many others what is required is further integration of the two, where agricultural activities support livestock survival, such that less pressure is put on the open range. In any case the present state of rural administration is not adequate, either in its geographical and social proximity to the people or in the resources it has available, to undertake any such programs.

Ultimately, the nature of the state – which has been hinted at but not analyzed in depth in this chapter, for lack of space – will determine the possibility of arriving at an appropriate and equitable distribution and use of the nation's resources. The changes required in the political and administrative arrangements constituting legal authority are the most vital aspect of any genuine attempt to combat desertification and reduce the impact of the inevitable droughts.

References

Abakr, A. R. and D. Pool 1980. The development process in its political-administration context. In *Problems of savannah development: the Sudan case,* G. Haaland (ed.). Occas. Pap., no. 19. Department of Social Anthropology, University of Bergen.

Abu Sin, M. H. 1980. Nyala: a study in rapid urban growth. In *Urbanization and urban life in the Sudan*, V. Pons (ed.). Development Studies Research Center, University of Khartoum, and Department of Sociology and Social Anthropology, University of Hull.

Adams, M. 1982. The Baggara problem: attempts at modern change in Southern Darfur and Southern Kordofan. *Develop. Change* **13**, 259–90.

Adams, M. and J. Howell 1979. Developing the traditional sector in the Sudan. *Econ. Develop. Cult. Change* **27**, 505–18.

van Apeldoorn, G. J. 1981. *Perspectives on drought and famine in Nigeria.* London: George Allen & Unwin.

el-Arifi, S. 1979. Some aspects of local government and environmental management in the Sudan. In *Proceedings of the Khartoum Workshop on Arid Lands Management,* J. Mabbutt (ed.). Tokyo: United Nations University.

el-Arifi, S. 1980. The nature and rate of urbanization in the Sudan. In *Urbanization and urban life in the Sudan,* V. Pons (ed.). Development Studies Research Center, University of Khartoum, and Department of Sociology and Social Anthropology, University of Hull.

Baker, R. 1976. The administrative trap. *Ecologist* **6**, 247–51.

Barnett, A. 1977. *The Gezira Scheme: illusion of development?* London: Frank Cass.

Barnett, A. 1981. Evaluating the Gezira Scheme: Pandora's box or black box? In *Rural development in tropical Africa*, J. Heyer, P. Roberts and G. Williams (eds). London: Macmillan.

al-Bashir, H. 1980. *Mechanized farming in Habila.* Paper to Conference on Regional Development, Kadugli, Sudan.

Bryant, M. 1977. Bread-basket or dust-bowl? *Sudanow*, October.

Digernes, H. T. 1977. *Wood for fuel: energy crisis implying desertification: the case of Bara, Sudan*. Cand. Polit. thesis. University of Bergen.

Digernes, H. T. 1978. *Fuelwood crisis causing unfortunate land use – and the other way round*. Paper for 8th World Forestry Congress, Jakarta, Indonesia.

el-Din, S., A. G. Undated work. *The gum tree and land use practice.* Unpublished.

Due, J. M. 1980. A note on agricultural credit in Sudan. *Savings Develop*. **4**, 219–33.

Duffield, M. 1979. *Hausa and Fulani settlement and the development of capitalism in Sudan*. PhD thesis, University of Birmingham (UK).

el-Faki, M. M. H. 1977. *Physical and socio-economic aspects of desertification in the northern part of Northern Kordofan Province*. Diploma of the Physical Environment in the Tropics. University of Khartoum.

Gaitskell, A. 1959. *Gezira: a story of development in the Sudan*. London: Frank Cass.

Garcia, R. V. 1981. *Drought and man: the 1972 case history*. Vol. I: *Nation pleads not guilty*. Oxford: Pergamon.

Glentworth, G. and M. S. Idris (eds) 1976. *Local government and development in the Sudan: the experience of Southern Darfur Province,* vols I and II. Development Administration Group, University of Birmingham (UK) and Sudan Academy of Administrative and Professional Sciences.
Government of the Sudan 1976. *Desert encroachment control and rehabilitation program.* Ministry of Agriculture, National Council for Research, and Food and Agriculture Organization, Khartoum.

Haaland, G. 1977. Pastoral systems of production: the Bon's cultural contract and some economic and ecological implications. In *Land use and development,* P. O'Kedefe and B. Wisner (eds), 179–93. London: International African Institute.
Haaland, G. 1980. Social organization and ecological pressure. In *Problems of savannah development: the Sudan case,* G. Haaland (ed.). Pap., no. 19. Department of Social Anthropology, University of Bergen.
Hales, J. M. Undated work. *The pastoral system of the Meidob.* PhD thesis. University of Cambridge.
Harrison, M. N. 1955. *Report on a grazing survey of the Sudan.* Ministry of Animal Resources. Cyclostyled.
el-Hassan, A. M. 1981. *The environmental consequences of open grazing in the Central Butana, Sudan.* Environ. Monogr. Ser., no. 1. Institute of Environmental Studies, University of Khartoum.
Hunting Technical Services 1974. *Southern Darfur land use plan.* London: Hunting Technical Services.
Hunting Technical Services 1976. *Savannah development phase II.* Rome: Food and Agriculture Organization.

Ibrahim, F. N. 1980. *The processes of desertification in rainfed cultivation areas.* Conference on desertification, Araak Hotel, Khartoum.
Ibrahim, F. N. 1981. Role of women in desertification. In *Women and the environment,* D. Baxter (ed.). Environ. Res. Pap. Ser., no. 2. Institute of Environmental Studies, University of Khartoum.
Ibrahim, I. M. 1982. *Trends and issues in local government finance in the Sudan.* M. Soc. Sci. dissertation, University of Birmingham (UK).
Ibrahim, S. A. 1981. Women's role in deforestation. In *Women and the environment,* D. Baxter (ed.). Environ. Res. Pap. Ser., no. 2. Institute of Environmental Studies, University of Khartoum.
International Bank for Reconstruction and Development 1966. *The Gezira Scheme.* Mission Report Annex. Washington, DC: IBRD.

Jefferson, J. H. K. 1955. *Soil conservation in the Sudan, development and projects.* Khartoum: McCargwadale.

Karam, K. M. 1981. *Dispute settlement among pastoral nomads in the Sudan.* M. Soc. Sci. thesis, University of Birmingham (UK).
Khogali, M. 1979. Nomads and their sedentarisation in the Sudan. In *Proceedings of the Khartoum Workshop on Arid Lands Management,* J. Mabbut (ed.). Tokyo: United Nations University.

Manger, L. 1981. *The sand swallows our land.* Bergen occasional papers in Social Anthropology, no. 24. Department of Social Anthropology, University of Bergen.
Meillassoux, C. 1974. Development or exploitation: is the Sahel famine good business? *Rev. Afr. Polit. Econ.,* no. 1, 27–33.
Ministry of Agriculture 1944. *Report of the Soil Conservation Committee.* Khartoum.
Ministry of Agriculture 1977. *Yearbook of agricultural statistics.* Khartoum.

Ministry of Agriculture. Undated report. *Working party's report on the proposed MCPS extension, South Gedaref District,* G. Hassan (Chairman). Khartoum.

O'Brien J. J. 1977. *How traditional is traditional agriculture in the Sudan?* Bull., no. 62. Economic and Social Research Council, Khartoum.

O'Brien, J. J. 1980. *Agricultural labors and development in the Sudan.* PhD thesis. University of Connecticut.

Pearson, M. 1980. *Settlement of pastoral nomads: a case study of the New Halfa irrigation scheme.* Develop. Stud. Occas. Paps, no. 5. University of East Anglia.

Pons, V. (ed.) 1980. *Urbanization and urban life in the Sudan.* Development Studies Research Center, University of Khartoum, and Department of Sociology and Social Anthropology, University of Hull.

Salam, M. M. A. 1979. *The institutional development of the Gezira Scheme.* PhD thesis. University of Reading.

Saeed, M. H. 1980. Economic effects of agricultural mechanization in rural Sudan: the case of Habila, Southern Kordofan. In *Problems of savannah development: the Sudan case,* G. Haaland (ed.). Occas. Pap., no. 19. Department of Social Anthropology, University of Bergen.

Sheets, H. and R. Morris 1976. Disaster in the desert. In *The politics of natural disaster: the case of the Sahel drought,* M. H. Glantz (ed.). New York: The African Studies Association.

Shepherd, A. W. 1982a. *Agricultural capitalism and rural development in the Sudan.* Paper to Development Studies Association (UK) Conference, Dublin, September.

Shepherd, A. W. Forthcoming. *Water, pastoralism and agricultural schemes: a case study from Eastern Sudan.* Occas. Pap., no. 17. Development Administration Group, University of Birmingham (UK).

Shepherd, A. W. 1983. Capitalist agriculture in the Sudan's *dura* prairies. *Develop. Change* **14**, no. 2, 297–320.

Shepherd, A. W. and A. M. el-Neima 1982. *Popular participation in decentralized water supply planning: a case study of Western District, Northern Kordofan.* Working Paper. International Labor Office, World Employment Programs.

Simpson, I. G. and M. C. Simpson 1978. *Alternation strategies for agricultural development in the central rainlands of the Sudan.* Rural Develop. Study, no. 3. University of Leeds.

Sørbø, G. M. 1977. Nomads on the scheme – a study of irrigated agriculture and pastoralism in Eastern Sudan. In *Land use and development.* P. O'Kedefe and B. Wisner (eds). London: International African Institute.

el-Tayeb, S. A. 1981. *The impact of water points on environmental degradation: a case study of Eastern Kordofan, Sudan.* Environ. Monogr. Ser., no. 2. Institute of Environmental Studies, University of Khartoum.

Thimm, H. U. 1979. *Development projects in the Sudan: an analysis of their reports with implications for research and training in arid land management.* Tokyo: United Nations University.

5 *Differential development in Machakos District, Kenya*

MARILYN SILBERFEIN

In the territory of what is now Machakos District, Kenya, a complex system of survival was devised by the Kamba based on agricultural techniques that had evolved through trial and error, symbiotic relationships between the occupants of different ecological zones, and a high level of mobility. This system was disrupted by the advent of the colonial period in the late 1900s. Boundaries began to be drawn around 'tribal' areas, interfering with the ability of people, and then cattle, to relocate in response to changes in weather or the population to land ratio. Competition was fostered in place of the Kamba traditions of cooperation in carrying out of agricultural tasks and mutual assistance during periods of crisis. The colonial administration emphasized commercialism in those areas that were deemed suitable for cash crops, at the same time that a policy of neglect was adopted toward other areas that were considered too isolated or too dry for economic development.

As the colonial regime gradually undermined the established system of survival they attempted to substitute their own strategies, but these imposed techniques were either poorly communicated or inappropriate given the resource base of the typical Kamba farmer. In effect, as population increased, colonial policies could not cope with land deterioration and frequent famines. Little progress could be made until after independence when the importance of both the neglected lowlands of Machakos District and of the established Kamba agricultural system were recognized. At this point the best of the Kamba strategies could be amalgamated with elements of the superimposed technology of the colonial and postcolonial period.

Initial conditions

The focal area of this study, Machakos District, Kenya, is a composite of highlands and lowlands (Fig. 5.1). Granitic hill massifs reach an elevation of over 7000 ft and the general height of basin varies between 3000 and 4000 ft. The hill slopes were still covered primarily by forests in the 19 century, but much of the agriculturally desirable land had been stripped of tree growth by the beginning of the colonial period. In the low lying areas a mixture of thornbush, herbaceous plants, and scattered trees has come to predominate, but the thornbush (often infested with tsetse fly) has been pushed back whenever cattle grazing or cultivation was extended onto the lowlands. Thus, the frontier between the varying vegetation types has frequently changed position with the vicissitudes of local conflict, rainfall variability, changing population densities and, ultimately, the colonial influence.

The Kamba have been gradually spreading through this territory since entering the area from a more southerly source region, probably in the early

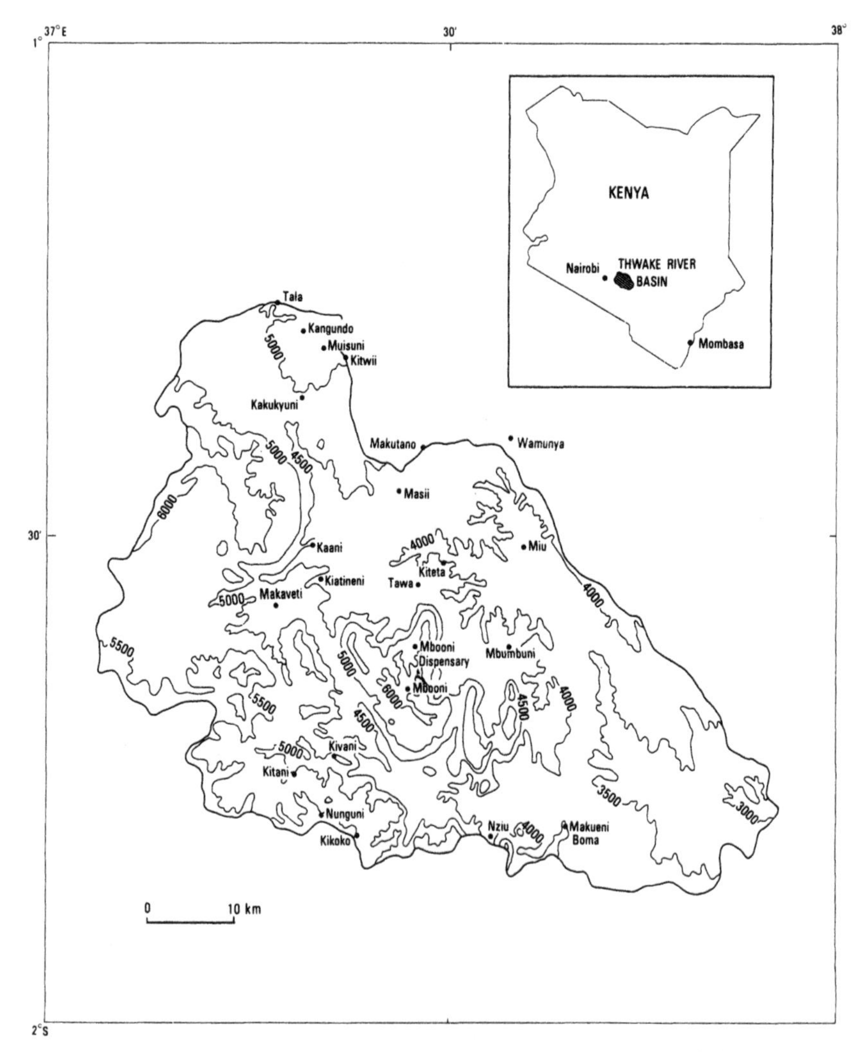

Figure 5.1 Thwake River Basin: sixth order towns in Machakos District. (*Sources:* British War Office 1953, Survey of Kenya 1965.)

1600s. The first site of Kamba concentration within the district was the Mbooni Hills (between 1650 and 1750), which provided protection from any external threat as well as a propitious setting for agricultural production. Movement out of Mbooni in the 18th century may have resulted from a combination of population growth and the reassertion of a more pastoral life-style that required extensive grazing land (Jackson 1977, p. 202).

At the same time that the Kamba were dispersing horizontally, they were oscillating between the higher and lower slopes of the occupied hill masses. Homesteads were usually located at higher elevations where both a dependable moisture supply and security from such competitors as the Masai were readily available. During the day, individuals moved down slope to herd cattle or to cultivate some of the moist depressions near stream courses.

The mobility of the Kamba has at least in part been a response to the vagaries of the rainfall regime. Machakos town receives an average of 32 in

Table 5.1 Average annual precipitation in selected areas of Machakos District, 1943–60.

	High rainfall[a] (in)	High–medium rainfall[b] (in)	Low rainfall[c] average (in)
Data from:	High areas: Machakos, Mbooni, Kilungu, Ngelani, and Matungulu	Low areas: Matiliku and Okia	Makuani, Meu, Sisthani, Maknvote, and Kiteta
1943	23	23	21
1944	35	28	21
1945	26	27	24
1946	38	30	26
1947	48	36	35
1948	42	30	26
1949	20	21	16
1950	25	20	19
1951	62	70	53
1952	31	28	17
1953	31	27	26
1954	36	29	26
1955	34	31	28
1956	32	30	26
1957	44	40	35
1958	37	32	27
1959	28	25	18
1960	27	20	16
average	34	30	25

Source: Peberdy.

of rainfall divided between a primary maximum in October–December and a secondary maximum in March–April. There is substantial variation in rainfall, however, from one year to the next, as well as from the hills to the plains (Table 5.1). Even more important than the total amount of precipitation is the distribution of rainfall through the growing season. When rainfall is received in infrequent heavy bursts, as is often the case, there is excessive runoff and erosion. A further consequence of the alternation between short, intensive moist periods followed by longer dry phases is an inadequate moisture supply for plant growth. In order to survive periods of erratic rainfall, the Kamba developed adjustments which were either partially or completely implemented, depending on the severity of the water shortfall. These adjustments can be grouped into five spheres of activity: (a) the agricultural system proper, (b) ritual practices, (c) the establishment of interdependencies, (d) participation in supplemental, non-agricultural pursuits, and (e) the use of staggered mobility. Each of these will be discussed in turn.

The agricultural continuum At the beginning of virtually every planting season farmers would utilize such practices as planting drought-resistant crops early whereas others were staggered over the rainy season. It was thus possible to salvage some harvest in the event of unpredictable, late, or irregular precipitation. Increased emphasis was placed on these techniques whenever rainfall totals fell below the average for several years in succession. Additional activities were also initiated, such as an increase in the total area planted or more complete weeding to cut down evapotranspiration. Efforts to obtain and control scarce water supplies were expanded – including well digging in dry river beds. Cattle would be increasingly relied on as a direct source of emergency foodstuffs.

The ritual continuum A complex sequence of rituals dealing with the supernatural gradually evolved in Ukambani as one mode of encouraging satisfactory production levels. The natural landscape was seen to be filled with symbols and signs: a rainbow, for example, might mean that no rain or at most a small quantity was forthcoming. Sacrifices, usually of chickens, were made in each community at designated natural sites known as *Ithembo*.

These ceremonies became more crucial with precipitation shortfalls since the failure of the rains was associated with the displeasure of the supernatural. The most substantial *Ithembo* would draw increased attendance during drought years, and larger animals were included in the sacrifices.

The interdependence continuum During normal years dispersed Kamba clans maintained close associations through dry season visitation. Other links were established through brotherhoods and both small- and large-scale trading networks. Since the Kamba occupied varied ecological zones, they were able to take advantage of variations in production over short distances. They might, for example, exchange fruits or vegetables produced in the hills for cattle products derived from lowland pastures. These trade associations as well as family-based relationships could be activated in time of stress. The tradition of general cooperation and sharing of scarce resources facilitated assistance to the hardest hit Kamba during a drought.

Supplemental activities Although many Kamba engaged in nonagricultural activities during years of average or normal rainfall, viz. hunting and trading, the involvement of individual farmers in these kinds of pursuits increased markedly in times of environmental stress. Other economic adjustments included the increased application of labor to the production of bricks and charcoal, and the sale of cattle for emergency income. Also important was the collecting of natural products to substitute for missing food items in the daily diet: fish, fruit, berries, and roots that were not usually eaten under normal conditions became important dietary supplements.

The mobility continuum The Kamba were known for their capacity to move over long distances for the purpose of colonization and trade, but their mobility also provided a mechanism for surviving water shortfalls. The most basic form of this strategy was the practice of field dispersals. By maintaining plots at several elevations, the cultivator insured that below normal yields might be balanced by greater productivity on the moister, higher slopes. This practice might be extended to include a satellite farm, located at some distance from the main fields so that daily visits were quite impractical. A single family member would maintain the outpost during the peak labor period and could thus provide a supplemental food supply in the event that the primary holdings failed to produce a normal crop. It was also common to keep cattle on the lower elevation farms while hillside plots were utilized for crops.

When precipitation declined below average levels, the Kamba were forced to implement additional components of their dispersion strategy. Initially, more males might become involved in hunting or gathering activities in order to generate income. In a more difficult situation, children might be sent away while parents attempted to salvage a meager harvest. Relatives living in better watered areas were most likely to receive the children, but even such neighbouring groups as the Kikuyu might accept limited numbers of outsiders who were willing to work in exchange for foodstuffs.

Under the most extreme conditions, whole families were forced to relocate, sometimes over long distances. As a rule such movements were temporary, but they became permanent if subsequent obstacles made it difficult to return to Machakos. Several small groups of Kamba founded settlements in north-central Tanzania when their return route was blocked by roving Masai; settlements that remain intact to this day.

The establishment of Machakos District, Kenya Colony

The International British East Africa Company (IBEAC) organized the first long-term intrusion into the Machakos area in 1889 when they established a supply station at what is now Machakos town for their planned foray into Uganda. The influence of IBEAC never extended very far beyond the site of the town, however. It was not until it became clear that company government was to give way to some form of colonial administration that an attempt was made to establish British authority elsewhere in the district (Forbes-Munro 1975, p. 29).

With the beginning of the British protectorate in 1895, the regional commissioner moved to eliminate slave and cattle raiding while expanding the British zone of control: first to the northern frontier area where additional posts were built, and then later to the south (Goldsmith 1955, p. 32). When groups of Kamba responded negatively to this display of authority, burning British posts and engaging in other acts that were deemed to be hostile, punitive expeditions were sent out to stifle the revolt.

The period of primary Kamba resistance ended in 1898 (Forbes-Munro 1975, p. 46). Several factors had undermined their capacity to withstand the expanding British influence and control. A severe famine during 1898–9, complicated by disease, rinderpest, and locust attacks, rendered the local population particularly vulnerable. Furthermore, some areas were beyond the range of these early encounters, and others had well established trading ties with the colonial government and remained neutral throughout the period of conflict.

When the military situation became less critical, colonial policies were devised and implemented among the reluctant Kamba that were to create long-range problems for the district. Taxes were introduced in 1901 and shortly thereafter the district was divided into locations, most of which corresponded to a distinct physical region such as hill massif as well as to an average of six distinct communities.

Simultaneously, land was being opened to European settlement. The Crown Lands Ordinance of 1901 included a plan for 99 year leases, which were to be expanded to 999 years in 1915. By 1906 the Athi and Kapiti Plains west of Machakos had been set aside for British settlers and the Yatta Plateau to the east was similarly restricted. Between 1908 and 1910 a further alienation came under discussion – that of the Mua Hills which, according to later surveys, formed a substantial component of the high potential land available to the Kamba.

The excuse given for land alienation was the sparse population of the district. Reports of missionaries and travelers during the late 1800s allude to a paucity of occupants in such areas as Momandu (Kalama) and Matungulu (Johnston 1899). According to another commentary, 'areas cultivated were small and much of the wood cut down was replaced naturally' (Stuart Watt 1899–1900).

These observations reflected the limited perspective of the outsider as much as the reality of conditions in Ukambani. To those unfamiliar with the agricultural system and the cycle of cultivation and temporary fallow, the territory available to the Kamba must have seemed adequate; certainly there was little evidence of erosion along slopes or sediment clogging rivers. This impression was artificially reinforced by two factors: (a) population losses and migration during the famine of 1899–1900 caused densities to decline in selected areas, and (b) too many of the estimations of the population–land balance were taken either in lowland areas that served as a spillover pasturage for cattle or in the frontier zones of both northern and southern Machakos which were still partly occupied by bush and game. Certainly, the most discerning observers mention either that it was difficult to estimate population densities since fields were scattered and dwellings often hidden (Watt 1900, p. 270) or that some of the hill massifs were fairly thickly populated (Ainesworth 1905, p. 411). Ainesworth's estimates for the Iveti Hills, for example, were 150 persons per sq. mile (see Gregory 1896, p. 347).

In effect, land alienation was based on an inaccurate view of the current situation and little concern with the impact of either population growth or precipitation shortfalls on the future territorial needs of the Kamba. Yet, population increases had become evident by the time the first settlers began to arrive. It is difficult to reconstruct precise figures, but relative population density can be estimated from the censuses that were taken at irregular intervals beginning in the early 1900s. These population counts, based on some combination of tax register information, number of dwelling units, and educated guesses as to the average family size, were almost always below the actual figures. The estimated population of Machakos District gleaned from the political record book and annual reports rose from 102 000 in 1902 to 130 000 in 1910. When a more careful census was taken in 1918, it showed a district population of 138 400 and some strong regional variations in density and physical conditions.

After 1910, the government focused on encouraging two groups within the population: (a) potential laborers who would move in and out of European farms as their services were required, and (b) subsistence producers who would grow enough to sustain themselves during a drought. Africans were forced into one of these categories because Europeans controlled the marketing boards, monopolized much of the best land, and dominated access to agricultural extension, inputs such as fertilizer or insecticide, and capital goods (Leys 1974, p. 34). The perfunctory effort made by the administration in the sphere of African agriculture did not include encouraging cash crops unless they contributed to the needs of the settler community.

Although the government achieved its goal of creating a mobile work force, its secondary goal, preventing famines, proved to be more elusive. A severe drought occurred in the district during the period 1898–9 and another followed in 1909–10 (Goldsmith 1955, pp. 46–7). The government instituted famine relief stations but these efforts would have been inadequate were it not for the retention of established survival mechanisms among the Kamba.

The Kamba techniques for adapting to drought did not remain unchallenged, however. Ritual activities were beginning to be undermined as missionaries spread an alternative belief system, and *Ithembo* sites were desecrated. The mobility strategy was even more seriously threatened. The Kamba pasturalists required relief grazing to supplement their inadequate pasture in 1909, but no territory was made available for 2 years. The occupants of the Iveti Hills were also forced to cross the boundary of the alienated Mua Hills in search of pasturage and firewood (Government of Kenya 1911–14).

An overview of the early colonial period in East Africa reveals a government lacking in sensitivity to the needs of the African population. No effort was made before Word War I to assess the current or future land requirements of the original occupants of Machakos District; instead wage-labor was put forth as an alternative to grazing cattle and farming the land. When they were faced with rebellion and later civil disobedience over the boundary issues, the authorities reacted with forced removals of 'trespassing' Africans and the encouragement of stock sales. Instead of formulating a coherent approach to the threat of famine, they substituted a limited endorsement of drought-resistant crops and an inadequate program of famine relief.

The interwar years

When the First World War ended, Machakos District received a temporary agricultural officer but the district remained in a marginal position from the purview of the colonial government. The administration was barely aware of the increasing participation of the Kamba in the cash economy and their growing enrollment in the regional educational system. Gradually, new shops were being established by both Indian and Kamba merchants and the number of registered public markets rose to 11 in 1929 (Forbes-Munro 1975, p. 171).

Government officials only began to take an active interest in African smallholders when, in the midst of the depression, increased productivity on African holdings came to be viewed as a potential source of much needed revenue. During the 1930s new crops were introduced and demonstration farms were established to exhibit the possibilities of mixed agriculture on small farms (Government of Kenya 1932, p. 2). These efforts were not particularly successful: infrastructure was inadequate and newly appointed extension agents, poorly versed in the details of local farming systems, could not integrate this established system with unfamiliar methods and crops. Furthermore, the extension staff was often unable to communicate with the mass of the Kamba peasantry, devising messages best suited to the few farmers who were completely integrated into the commercial sector.

It was during this phase of increased government involvement in African agriculture that the policy of subregional favoritism emerged. Attention was focused on the highlands which had exhibited a relatively positive reaction to cash cropping, whereas the lowlands were neglected. The government, in effect, initiated a self-fulfilling policy: the highlands produced the most impressive results in terms of registered farmers, acres in cash crops, new cash crops cultivated, and highest yields, but the accomplishments of the highland farmers were to a great extent due to better roads, better marketing facilities, and an unequal share of the agricultural inputs available to the district.

Meanwhile the lowlands, viewed as a problem by the government, fell further behind. The drier areas were administered through somewhat of a holding action. The negative effects of droughts were minimized, but almost no activist policy was pursued that would have improved the commercial status of both livestock keeping and agrarian-based activities. A statement made in the Agricultural Report of 1938 sums up this attitude: 'Yields are improving in the higher areas which are more fertile but in other areas it is difficult to see much improvement and the soil is becoming even more impoverished' (Government of Kenya 1938, p. 9).

The emergence of the important highland–lowland dichotomy was also reflected in the distribution of missions and schools, which in turn provided alternative modes for the diffusion of agricultural knowledge. The first permanent missions were established in the vicinity of Machakos town. Once these sites were occupied, the next locations to be chosen were in the highlands where an equitable temperature regime and a presumption of high agricultural potential were combined with a relatively large number of potential converts to Christianity. A few missions were founded just beyond the hill massifs of the edge of the lowlands (Mukaa, for example), but for a long time much of the southern and eastern parts of the district was uninfluenced by either the direct or indirect ramifications of Christian prosyletizing (Forbes-Munro 1975, p. 121).

The government followed the missions' lead by encouraging primary education in bush schools. Almost all of these schools were within or close to the highlands in spite of attempts to achieve district-wide coverage. According to government commentaries, schools were to be built in specific lowland communities but there was no real effort to produce the financial assistance needed for this particular plan. As late as 1927, when 21 additional bush schools were being designated, the new sites were once again located predominantly in the highlands.

Also contributing to the highland–lowland dichotomy was an economic focus on agrarian rather than pastoral activities, a situation best suited to the better watered highlands where fruits, vegetables, and wattle could be grown. The price of maize nearly doubled during the 1920s whereas the price of hides actually fell (Van Zwanenberg 1975). Cattle prices in the early 1920s to the late 1930s fluctuated considerably but a downward trend was definitely established. The problem was exacerbated by the tendency for cattle sales to reach peak levels at times of drought when prices were likely to be low.

The differences between lowland and upland continued to be intensified by the prejudices of the agricultural administration. In 1932, the following statement was made concerning the introduction of new seed as well as improved maize varieties: 'work in the district has concentrated on the more advanced areas but it is hoped to get to the arid areas shortly' (Fig. 5.2a). The 1934 Agricultural Report contained the remark that increasing progress was being made in the highlands, and in 1935 the agricultural office took the position that only selected areas would be worth a heavy investment of personnel (Fig. 5.2b). As for the drier areas, there was little expectation of success because of 'lower rainfall, unresponsiveness, and the relatively larger holdings that were deemed necessary for each farm unit' (Government of Kenya 1937, p. 12).

The combined effect of all these factors was the creation of two very different sets of social and economic conditions. While the highlands prospered, the south and east fell into an increasingly backward role. The problem was recognized by some administrators as evidenced by the following statement: 'It is hoped next year to open plots in arid areas which are much in need of early maturing and drought resistant varieties of crops' (Government of Kenya 1932). Yet, only one major effort was made to introduce a new crop, an experiment with cotton planting in the 1930s. Almost 3000 acres were eventually planted, but yields remained low, undermining the initial enthusiasm for a crop that 'provides a useful contract between the government and the more remote parts of the district and without which many farmers will have no cash' (Government of Kenya 1940–3). The effort was eroded by compulsion, low market prices, pest problems, and drought, and was finally abandoned without a sufficiently long trial period.

By concentrating on commercial agriculture in the highlands the administration was, in effect, ignoring pervasive problems: growing land

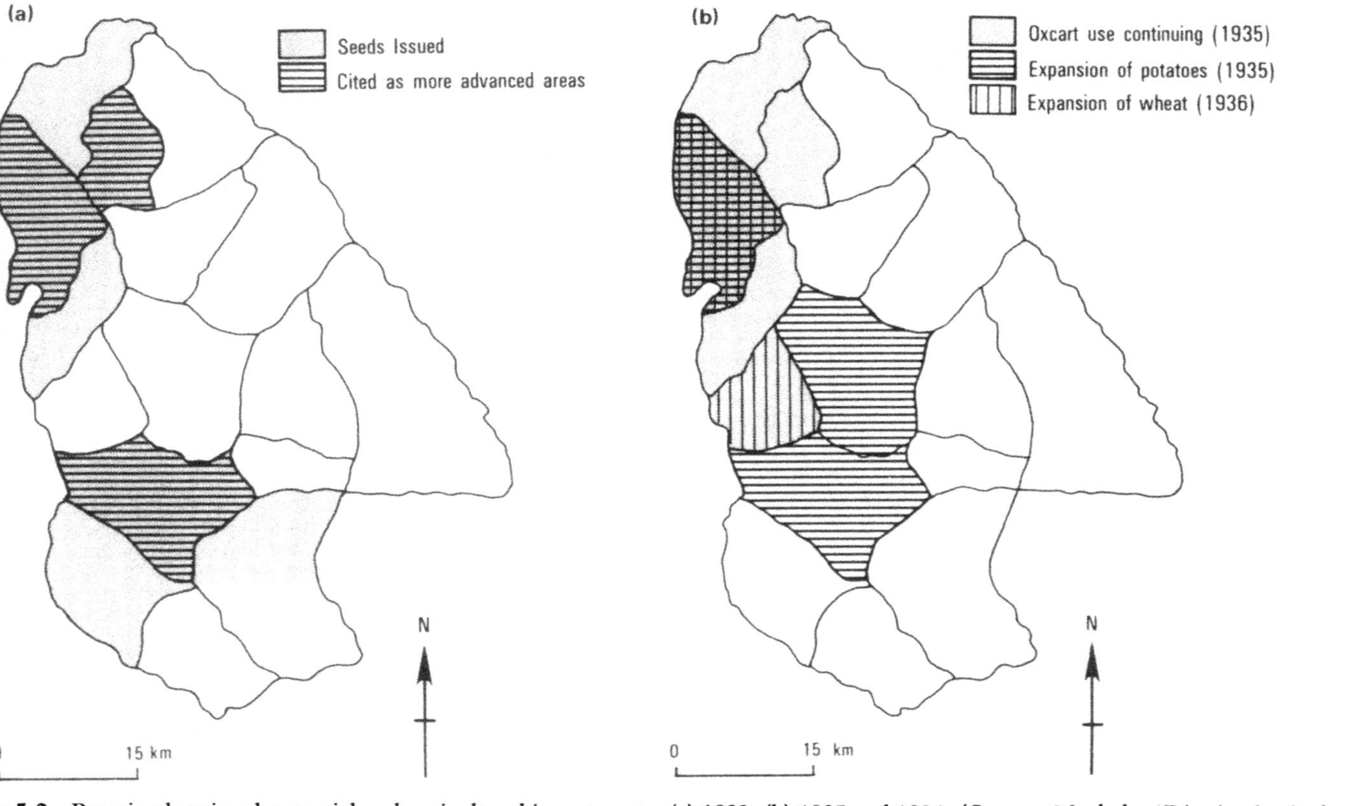

Figure 5.2 Perceived regional potential and agricultural investments: (a) 1932; (b) 1935 and 1936. (*Sources:* Machakos District Agricultural Reports 1932, 1935, and 1936.)

scarcity due to population increases combined with such manifestations of land deterioration as slope erosion. The 1939 population estimates showed increasing concentrations in the highlands, especially in the north, while the lowlands exhibited densities of less than 200 persons per sq. mile (Fig. 5.3a). These totals were somewhat deceptive as they masked more detailed patterns and failed to convey the precarious nature of the less than 150 per sq. mile densities in some of the lowland locations. By 1939 it was stated by the district officer that open unclaimed land could only be found in remote parts of the district where fly and bush still covered numerous tracts (Forbes-Munro 1975, p. 201). By 1948 some of the highlands had reached densities of over 800 persons per sq. mile for the first time (Fig. 5.3b). None of the locations could any longer be considered sparsely populated although, at the subdivisional level, small pockets of isolated or bush-covered land remained relatively empty.

Throughout this period of mounting stress Kamba survival strategies continued to operate, although at a reduced scale and often in a modified form. The mobility strategy was adopted to emphasize squatting – the exchange of labor on European farms for a small wage or the right to cultivate a subsistence plot and graze a few cattle. Other attempts to use the mobility strategy to deal with drought and population growth were frequently met with frustration. Before World War I migrants had begun to move into the lowlands in larger numbers. At first they advanced with their cattle in search of pasture while reserving a plot in the highlands; later they made a more permanent commitment to both arable and pastoral activities in the lowlands. The agricultural methods used by these migrants were extremely extensive and, in the delicate ecological setting of the lowland areas, land deterioration could set in rapidly. In the early 1920s, the annual reports comment on the depletion of grazing land and the need to drive herds into even drier areas and so 'reduce neighboring areas to an equal state of barrenness'. During the extremely dry years of 1924–5 many of these families were forced to move back into the highlands where land was already scarce.

A situation had been reached by the late 1930s whereby many Machakos District farmers were being forced to oscillate between highland and lowland areas with increasing restraints at either end. As land was becoming more of a scarce commodity, land sales, tenancy, and for the first time, a landless peasantry emerged as important phenomena within the district. Even when migrants reached out beyond the Machakos area, restrictions might be imposed by neighboring groups or by administrators seeking to avoid confrontations.

Other aspects of the traditional Kamba strategy became less significant. The expansion of missions and of commercial agriculture undermined the role of rituals and cooperative activities, respectively. Other practices changed in form but not necessarily in substance. Just as additional sources

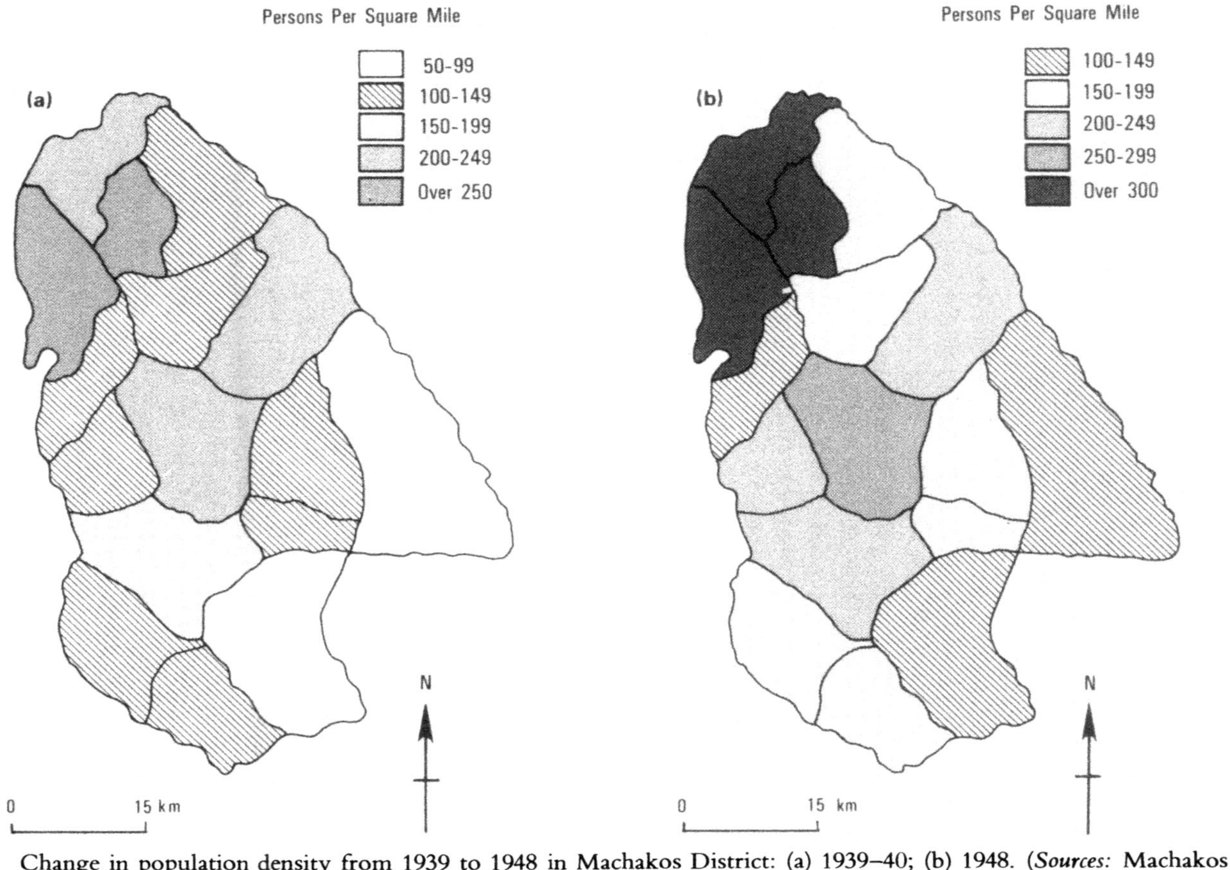

Figure 5.3 Change in population density from 1939 to 1948 in Machakos District: (a) 1939–40; (b) 1948. (*Sources:* Machakos District Agricultural Reports 1939 and 1940, Machakos District Census 1948.)

of income during periods of stress had once been sought from hunting and trade, so the Kamba of the interwar period looked to the production and sale of bricks and charcoal for the same purpose. The most important source of income was the sale of labor, and large numbers of Kamba, mostly young and male, attempted to earn wages in the cities and commercial farming enterprises of Kenya. The impact on village life of this type of migrant labor was much greater than during the period of hunting and long-distance trade, however, since more than 50% of the male inhabitants of a community might be absent at any one time.

When the government finally decided in the 1930s that a policy was required to deal with droughts, and food and land shortages, the indigenous Kamba approaches to coping with these issues were again ignored. According to the colonial administration, the Kamba were ruining their land with 'primitive techniques, and overstocking and timber cutting', and the only antidote was a combination of conservation measures and more extensive planting of famine-reserve crops (Government of Kenya 1925–6). A more realistic assessment would have taken into account the negative impact of the original land alienations and the Kamba attitude toward agriculture and herding. It was at this time, in fact, that the Kenya Land Commission was reviewing conditions in the areas that had been reserved for African farming. Although problems of increasing erosion and inadequate grazing land were obvious in the Kamba territory, the commission's members put the blame on the Kamba agricultural system and overgrazing rather than on the loss of half the established areas of pasturage to European settlers (Wisner 1976, p. 11).

Although the land question was deftly avoided, several policy lines emerged that were designed to halt the process of soil deterioration. These included selected reconditioning of the reserve and forced destocking. The reconditioning process which led to the spread of terracing and the temporary closure of selected areas to grazing also accelerated the emergence of individual land tenure. Although many Kamba had misgivings about this program, their reaction to destocking was much more negative. This hostility was derived from the government's initial presumption that the size of Kamba family herds was excessive. In actuality, the stock supply in Machakos District had not increased more rapidly than the population – many families had actually lost stock as a result of famines and a period of compulsory purchase during the war. The problem should have been couched in terms of Kamba needs with some reference to the importance of cattle in the local economy and the system of drought survival. The solution would have then been to increase the amount of land available. Instead, the emphasis was placed on reducing cattle numbers, and the provision of limited additional grazing land under tightly controlled conditions. This ill conceived project was finally postponed indefinitely following a Kamba protest march to Nairobi.

The impact of ALDEV and the Swynnerton Plan

The postwar years in Kenya were marked by the emergence of two programs designed to improve the productivity of the African rural sector. The first of these, the African Land Development Board (ALDEV), was formed in 1944 to amplify the theme of land reconditioning and related activities. The second program, known as the Swynnerton Plan when it was finally formalized in 1954, was geared to expanding African cash-crop production through improved markets and infrastructure, the distribution of appropriate inputs, and the gradual consolidation and enclosure of land holdings (Swynnerton 1955).

It was the latter plan that contributed to the further differentiation of highland and lowland Machakos. Both the prejudices of the agricultural staff on this issue, and the results of the unequal distribution of resources during earlier decades, continued to show up clearly in the agricultural reports. For example, in the 1950s the 'progressive areas' were defined as the highlands along with Mukaa and Masii and these same areas were identified as having the best diet, showing the most initiative in small-scale irrigation schemes, and making the greatest efforts at terracing (Government of Kenya 1957).

Given higher levels of government commitment, and an established commercial base, the highlands moved more definitely into cash-crop production after 1950. The restrictions that had prevented African farmers from growing such cash crops as coffee were finally lifted, in spite of continuing reluctance and fear of competition within the settler community. Coinciding with coffee expansion was the increasing importance of fruits and vegetables. Rainfed tomatoes, onions, cabbages, and cow peas were widely dispersed throughout the northern locations and Mbooni, as were such fruits as citrus, strawberries, peaches, and plums.

At the same time that the highlands were successfully adapted to an increasingly commercial agricultural economy, the larger farmers within the area were receiving a disproportionate share of the already generous amount of locally available inputs and expertise. It was during the 1950s that a distinct class of relatively wealthy farmers emerged. Many of these individuals resembled their settler counterparts in that they had clear title to their land, had acquired managerial skills, and depended extensively on hired labor (Datoo 1977, p. 72). They were also grouped into better farmers' clubs and asked to participate in classes at farmers' training centers (Ruthenberg 1966, p. 17). Their existence reflected one of the basic objectives of the Swynnerton Plan: 'to devote personal attention to individuals who were exhibiting progressive tendencies and, in turn, to create a group of yeoman farmers who would provide a stabilizing force in the rural area' (Government of Kenya 1952, p. 11).

The policy of encouraging selected farmers to enrich themselves brought about a polarization of what had been a fairly egalitarian society. It created differences in income and opportunity that could only be exacerbated by population growth and increasing commercialization (Leys 1974, p. 52). Gradually, a fluid situation in which farmers could expand their acreage or in other ways move into cash-crop marketing was changing into a more inflexible system that allocated inputs to a narrowly defined segment of the peasantry.

The lowlands during the postwar period were still considered barely suitable for the expansion of commercial agriculture. The idea of reintroducing cotton was discussed in the 1951 Agricultural Report but the general concensus was negative. The area was described as still primarily subsistence oriented, with a population that was apathetic and backward because 'no future is seen in the industrious working of the land which seems unable to give them a living' (Agricultural Report 1951).

As a result of this attitude, only a few cash crops were grudgingly introduced at lower elevations. An effort was made to investigate the commercial potential of fruits and vegetables that were already being grown in valley bottoms, but it was decided that marketing problems due to isolation or lack of organization precluded any expansion of production. In the late 1940s tobacco cultivation was attempted at a lowland settlement scheme but the project was abandoned, again because of marketing problems combined with a perception of inadequate fuel supplies (Owako 1969, p. 272). The District Council then encouraged sisal as an appropriate lowland crop and some success was experienced between 1946 and 1952. The potential popularity of sisal was reduced, however, by the 5 years required for its maturation and by an unstable market situation (Owako 1969, p. 260).

At this point, pastoral activities appeared to offer the best promise of a cash income in the lowlands although much more livestock control and related measures would have been required better to exploit the commercial potential of this resource. Dairying was proposed at one point but the familiar problem of erratic rainfall as well as the question of disease control put an end to this line of speculation. As a result of the limited sources of income available to lowland farmers, the economic differentiation of the highlands did not penetrate the southern and eastern portions of the district at this time. It was not yet obvious that the livestock industry, still relatively disorganized, would become a source of wealth to the herders who would be able to acquire enough land to establish cattle ranches.

While the Swynnerton Plan expanded commercial agriculture, the ALDEV program continued to deal with the ongoing problem of precipitation shortfalls (Table 5.1). Serious food shortages which accompanied these precipitation crises had been occurring on the average of once every 3 years and, as of 1943, parts of the lowlands had not been self-sufficient in food for over 10 years (Government of Kenya 1943). Some areas had experienced 2 or 3 years without sufficient rainfall, and a consequent lowering of the water table had made it difficult even to use sand wells (Mbithi 1971, p. 3). Stopgap measures such as government famine relief had been utilized, but it was obviously necessary to move away from emergency relief and substitute self-sustaining mechanisms that would ensure a continuing supply of foodstuffs for the district.

The specific ALDEV projects designed to alleviate these conditions were limited in scope. Farmers were encouraged to grow cassava and rapidly maturing maize but an administrative structure existed which would ensure that such plots were continuously cultivated. Soil conservation teams were set up based on groups of local farmers, and these groups restored some of the denuded Machakos landscape through grass planting and terracing. The building of dams was another line of defense against drought and the results between 1946 and 1962 were initially impressive. Unfortunately, lack of maintenance led to some dams silting up and spillways became gullied from heavy grazing.

Another ALDEV project with mixed results was the establishment of lowland settlement schemes as a means of relieving overcrowded areas and of providing land for unemployed urbanites. Most of these resettlement efforts were plagued by cost overruns, a dearth of suitable cash crops, a less than ideal settler population, inadequate or inappropriate personnel and, ultimately, a level of production which was not appreciably different from neighboring areas that had not had the advantage of government investment (Ruthenberg 1966, p. 56). Several thousand settlers eventually moved into the major scheme in southern Machakos District, Makueni, and others were able to make use of a 240 000 acre grazing scheme.

Toward independence and beyond

Agriculture policy as it applied directly to the highlands did not change direction in the early 1960s. The general staff of the Ministry of Agriculture continued to be concentrated in these areas as were the specialized services and the statutory boards (Heyer *et al*. 1971, p. 47). The favored cash crops were the focus of attention, with coffee becoming even more important.

The special position of the highlands was approved of in the International Bank Report which claimed that given current technologies, agrarian development ought to be limited to specific areas, such as the highlands (International Bank for Reconstruction and Development 1963, p. 64). In contrast, the World Bank's recommendations for the lowlands emphasized livestock unless irrigation facilities were available. Other evaluations confirmed the prognosis that low and uncertain rainfall combined with medium to poor quality soils was not going to be conducive to successful commercial agriculture (DeWilde 1967, p. 86).

It would seem that the World Bank Report was dated before it was even published, because by 1963 population density in the lowlands had risen to a level that was above the ideal for ranching. Since alternatives were needed, cotton was experimentally reintroduced in 1961. It was rapidly taken up by farmers anxious for a source of income but yields tended to remain low (an average of 350 lb per acre as opposed to the estimated potential of 1500) (DeWilde 1967, p. 109). The major reason for the poor level of cotton production was the labor requirement: the demands of this crop conflicted with those of maize which was usually grown twice each year. Farmers also had difficulties obtaining loans as the credit system was still highland oriented. Even when loan possibilities did exist, it was often the farmers with the greatest need, those with small, isolated holdings, who were the most hesitant to apply (Mbithi 1971, p. 226).

The actual arrival of independence in 1964 was marked by only one dramatic development in the agricultural sphere: the partial amalgamation of the European and African farming systems through African resettlement in the scheduled areas ('White' highlands) and through cooperative ranching. Machakos District itself was not a focus of government attention since the relatively successful programs of the 1950s combined with good weather in the early 1960s created a temporary illusion of progress and self-sufficiency (Lynam 1978, 61).

Gradually, new policies for rural development were introduced. A conference held in Kericho in 1966 produced a document citing the need for decentralization and the coordination of various activities at the local level. In order to facilitate these and related goals, the Special Rural Development Program was created in the early 1970s and was charged with the design of six pilot projects in integrated rural developments. This approach would have undoubtedly been applied to numerous lowland sites, but it was discontinued because a concentration of capital and skilled managers on a limited number of projects came to be perceived as an inappropriate use of scarce resources (Heyer *et al.* 1976, 213).

Another rural development thrust, based on the search for rapidly maturing crops for semi-arid areas, led to the development of Katumani maize. This new variety was considered to be a possible long-term solution to the recurrent famine phenomenon but in spite of its obvious advantages, Katumani maize was not as widely adopted as had been expected. There were several reasons for farmer reluctance: most important was the association of Katumani with a package of activities that were necessary to insure the crops' success. Included in the package were proper planting and spacing, thorough weeding and a level of soil fertility that often called for the application of fertilizer (Hunt 1977, 84). Since many farmers could neither supply the labor needed for weeding nor afford to purchase fertilizer, they continued to rely on low-yielding local maize which would provide them with at least some returns. The first strains of Katumani maize were also not likely to do as well as regular maize during seasons that had above average rainfall totals (DeWilde 1967, 89).

These development efforts did not stop the disparity between highland and lowland from becoming more exaggerated. The vulnerable position of the lowlands intensified as a result of very rapid population growth. In some areas the growth rate had been in excess of 33% a year, 10 times the national average (Mbithi & Wisner 1972, 10). Many migrants in the area were exiles from the land-scarce highlands who lacked skills appropriate to dry land farming and cattle keeping. Others cultivated land that was unproductive during dry years and would better have been kept in pasture. Choice of crops and techniques were constrained because the need to guarantee subsistence in dry years was more important than maximizing potential income.

Another important development was the disappearance of ties that had previously enhanced the trade potential between the highlands and lowlands. The low lying areas had declined into a typically peripheral role, providing the highlands with labor and certain low-priced raw materials (cattle products in particular), and receiving relatively high-priced foodstuffs in exchange (Wisner 1977, 356). Other typical signs of a peripheral relationship were the low wages paid to migrant lowland laborers and the concentration of limited investment funds in highland-based projects where the immediate returns were expected to be higher (Porter 1979, 55). Thus, the relationship of these two regions had come to be characterized by exploitation rather than by mutually advantageous trade accompanied by sharing during periods of stress.

Just as reliance on assistance from kin and trade partners had dissipated by 1970, so had other aspects of the indigenous survival system. Alternative activities continued to be important to the lowlands during periods of stress, but with substantial limitations; since some of these activities require capital or specialized skills (running a shop, tailoring, carpentry) they were not widely available as survival strategies. Other options were closed off because they were based on access to certain scarce or expensive raw materials – as in the case of grain for beer making, or firewood for direct sale and charcoal production. Because of these inhibiting factors and as a result of the widespread availability of manufactured goods, numerous crafts and established skills were disappearing from the population.

The mobility strategy continued to play an important role among the Kamba but under very difficult circumstances. The spreading out of fields was undermined by land consolidation and registration, and satellite fields accessible to the home farm were frequently not available. As for the process of migration, it was also complicated by land scarcities and a growing tendency to invest in fixed capital that could not readily be replicated. Under these circumstances, a move by a family to a new location would entail the selection of a potentially marginal or illegal location, and a loss of resources.

By the early 1970s it had become obvious to the Kenyan government that activities to date had not solved the environmental problems, reduced highland–lowland disparities, or led to high levels of productivity and commercialism. The typical Machakos farm family in 1975 was still subsistence oriented and likely to consume at least 75% of its own production. Most labor was family based and seeds were usually obtained locally to include a residue from the previous year's harvest. This situation persisted because it was so difficult to increase productivity, diversity, or produce for the non-local market. Any of these adaptations carried with them too many constraints or too high a level of risk.

The most important constraint was that of capital shortages. The typical farmer had little income to purchase needed inputs and equipment or to engage workers during periods of peak labor demand. A viable credit structure would have eased this problem, but credit was usually most readily available to farmers who had already proven their record as commercial growers, or those who had some formal education.

Another constraint on farmers was lack of access to markets and services. In many areas, successful participants in the cash economy were confined to a strip 3–4 km wide on either side of all-weather roads. Compounding this problem were inadequate local markets at which produce could be sold or agricultural inputs obtained. An image can be conjured up of a typical smallholder carrying produce as a back load or head load to the nearest market located at 20 km from the homestead (Hash & Mbatha 1978, 51).

The years 1974–5 mark a turning point. The inadequacy of past policies had been accepted. The 'marginal areas syndrome' which had been allowed to develop in the lowlands was finally attributed as much to neglect as to local physical conditions or a fatalistic attitude among the inhabitants. The following statement is typical: 'Since much of the high potential land is now heavily populated, the government is having to consider more and more of the so-called marginal areas' (Dickson 1978, 21).

The third development plan (1974–8) was the first to pay serious attention to the lowlands. Negotiations began in 1975 with the European Economic Community (EEC) for a grant that would underwrite the first experiments with concentrated development in a semi-arid area – the Machakos Integrated Development Program (MIDP) (see Republic of Kenya 1977). The drier areas of Machakos District were to be developed through research into new crops that used moisture more effectively or technologies that allowed crops to maximize access to existing moisture supplies. Also important were conservation activities and integration of the lowlands into the mainstream of the national economy. There was an expectation that both additional employment opportunities and more equitable distribution of wealth could be achieved by improving productivity and the distribution of services (Government of Kenya 1977, 2–4).

The second aspect of the shift in government programs was a recognition of the efficacy of the Kamba economic systems. Attempts were finally made to integrate aspects of the Kamba agricultural system into the package of techniques espoused by the extension staff. For example, the mixed cultivation of such crops as maize, millet, and cow peas was recognized as a viable method for keeping down weeds, reducing the number of plant-specific pests, and allowing for a more thorough use of soil nutrients.

A further acknowledgement of the value of Kamba wisdom with regard to surviving in a difficult environment was the emphasis that the MIDP placed on participation. Rather than imposing a sequence of practices, the EEC program specifically solicited input from households in sublocations. This information was in turn filtered through local leaders, including chiefs, extension officers, and others, who would periodically meet together to examine and rank priorities. Town meetings (*barazas*) were also held to provide an opportunity for discussion (Myers 1981, 25).

Thus the Kenya government has finally been willing to re-evaluate its priorities. It is useful to view current efforts as restoring some of the initial conditions encountered by the British colonizers. The ideal economic system is now seen as one in which agriculture and animal husbandry are based in part on long-term experience and on an intimate knowledge of local environmental conditions, and the physiology and genetic variability of plants. At the same time, there is a recognition of the importance of internal regional interdependence and of development programs that take all subregions into account in terms of their distinctive qualities and potential.

References

Ainesworth, J. 1905. The tribes of the Ukamba, their history, customs, etc. *E. Africa Q.* **2**, 405–13.

British War Office 1953. Machakos sheet 1 : 250 000 GSGS 4672, 2nd edn.

Datoo, B. A. 1977. Peasant agricultural production in East Africa: the nature and consequences of dependence. *Antipode* **9**, 70–7.

DeWilde, J. C. 1967. *Experiences with agricultural development in tropical Africa*. Baltimore: Johns Hopkins University Press.

Dickson, M. 1978. Heavy emphasis on agriculture. *Finan. Times (Lond.)*, 21–3.

Forbes-Munro, J. 1975. *Colonial rule and the Kamba*. Oxford: Clarendon Press.

Goldsmith, F. H. 1955. *John Ainesworth, pioneer Kenya administrator*. London: Macmillan.

Government of Kenya 1911–14. *Political record book*.

Government of Kenya 1925–6. *Agricultural report*. Department of Agriculture.

Government of Kenya 1932. *Agricultural report*. Department of Agriculture.
Government of Kenya 1934. *Agricultural report*. Department of Agriculture.
Government of Kenya 1937. *Agricultural report*. Department of Agriculture.
Government of Kenya 1938. *Agricultural report*. Department of Agriculture.
Government of Kenya 1940–3. *Agricultural report*. Department of Agriculture.
Government of Kenya 1951. *Agricultural report*. Department of Agriculture.
Government of Kenya 1952. *Agricultural report*. Department of Agriculture.
Government of Kenya 1957. *Agricultural report*. Department of Agriculture.
Government of Kenya 1977. *Agricultural report*, 2–4. Department of Agriculture.
Gregory, J. W. 1896. *The great rift valley*. London: John Murray.

Hash, C. and B. Mbatha 1978. Economics. In *Kenya: marginal, semi-arid lands pre-investment inventory*, B. Palmer *et al.* (eds). Rep. no. 3. Ministry of Agriculture, Nairobi.
Heyer, J., D. Ireri and J. Morris 1971. *Rural development in Kenya*. Nairobi: East African Publishing House.
Heyer, J., J. K. Maitha and W. M. Senga 1976. *Agricultural development in Kenya: an economic assessment*. Nairobi: Oxford University Press.
Hunt, D. 1977. Poverty and agricultural development policy in a semi-arid area of Eastern Kenya. In *Land use and development*, P. O'Keefe and B. Wisner (eds). Afr. Envir. Spec. Rep., no. 5. London: University of London.

International Bank for Reconstruction and Development 1963. *The economic development of Kenya*. Baltimore: Johns Hopkins University Press.
International Bank for Reconstruction and Development 1975. *Kenya: into the second decade*. Baltimore: Johns Hopkins University Press.

Jackson, K. A. 1972. The dimensions of Kamba pre-colonial history. In *Kenya before 1900*, B. A. Ogot (ed.), 174–261. Nairobi: East African Publishing House.
Johnston 1899. *African Duland Mission report*.

Leys, C. 1974. *Underdevelopment in Kenya*. Berkeley: University of California Press.
Lynam, J. K. 1978. *An analysis of population growth, population change, and risk in peasant, semi-arid farming systems: a case study of Machakos District, Kenya*. PhD thesis. Stanford University. Unpublished.

Mbithi, P. M. 1971. *Famine crises and innovation: physical and social factors affecting new crop adoptions in the marginal farming areas of Eastern Kenya*. Rural Development Res. Pap., no. 52. Department of Rural Economy and Extension, Makerere University College.
Mbithi, P. M. and B. Wisner 1972. *Drought and famine in Kenya: magnitude and attempted solutions*. Disc. Pap., no. 144. Institute for Development Studies, University of Nairobi.
Myers, L. R. 1981. *Organization and administration of integrated rural development in semi-arid areas: the Machakos Integrated Development Program*. Agency for International Development, Office of Rural Development, and Development Administration.

Owako, F. N. 1969. *The Machakos problem: a study of some of the aspects of the agrarian problems of Machakos District, Kenya*. PhD thesis. University of London. Unpublished.

Porter, P. 1979. *Food and development in the semi-arid zone of East Africa*. Africa Ser., no. 32. Syracuse University.

Republic of Kenya 1977. *Machakos Integrated Development Programme: Submission to European Economic Community*. Nairobi: Ministry of Finance and Planning.

Ruthenberg, H. 1966. *Agricultural policy in Kenya, 1945–1965*. Berlin: Springer-Verlag.

Stuart Watt, J. A. Private papers. Rhodes House, Oxford.

Survey of Kenya 1965. Ed. 3–SK. Nairobi and Kitui sheets 1 : 250 000, Series Y503, SA-37-5, and SA-37-6.

Swynnerton, R. J. M. 1955. *A plan to intensify the development of African agriculture*. Nairobi: Government Printer.

Watt, R. S. 1900. *In the heart of savagedom*. London: Marshall.

Wisner, B. 1976. *Man-made famine in Eastern Kenya: the interrelationship of environment and development*. Disc. Pap., no. 96. Institute of Development Studies, University of Nairobi.

Wisner, B. 1977. Constriction of the livelihood system: the peasants of Tharaka Division, Meru District, Kenya. *Econ. Geog*. **53**, 353–7.

van Zwanenberg, R. 1975. *Colonial capitalism and labor in Kenya, 1919–1939*. Kampala: East African Literature Bureau.

6 *The demise of the moral economy: food and famine in a Sudano-Sahelian region in historical perspective**

MICHAEL J. WATTS

Famines do not occur, they are organized
by the grains trade *Bertolt Brecht*

It is now almost a decade since the beginning of a series of drought years which acted as a catalyst for the devastating famine throughout Sudano-Sahelian West Africa. The graphic portrayal of the inexorable southerly advance of the Saharan ergs, however preposterous, gained considerable credibility, fueled further no doubt by the more recent pessimism of the United Nations conference on desertification. The phenomenon of 'creeping deserts' and the ecological degradation of semi-arid lands has achieved a certain notoriety as is generally the case when an issue surfaces on the cover of *Time* magazine (September 12, 1977). Yet the empirical status of desertification is itself somewhat indeterminate, and its etiology a subject of some debate (Baker 1977, Kates *et al.* 1977).

A large part of the desertification–famine literature reflects the fascination with climatic instability, in particular the extreme spatial and temporal variability in Sahelian rainfall. Perhaps inevitably, in view of the paucity of data, there is no consensus of opinion among meteorologists: some favor long-term, secular changes in pressure conditions; others point to particulate matter, sunspots, climatic cyclicity, and albedo change. In view of the dissension that surrounds the climatic change issue, it is paradoxical that so much of the social scientific work on hunger and food shortage in arid West Africa should be couched rigidly in meteorological terms. This is particularly evident in the use of the term 'natural disaster' as a synonym for

*Part of this paper is an adaptation of an earlier manuscript that was jointly authored with Robert Shenton, University of Toronto. In addition, I would like to acknowledge the comments and criticisms of Gavin Williams, Bill Freund, Phil Porter, and Richard Palmer-Jones. The research on which this paper is based was financed by the following institutions: Social Science Research Council, the National Science Foundation, Resources for the Future, Inc., and the Wenner Gren Foundation for Anthropological Research. The ideas presented here are expanded on in Watts (1983).

famine, an equation that implies a direct, linear, and automatic relationship between unpredictable environmental changes and human starvation. Drought equals famine, and hunger assumes a 'no-fault' status, an Act of God to be written off like an insurance claim.

But clearly, to the extent that there is a relationship between famine and ecological stress or environmental uncertainty, it is mediated by the social, political, and economic constitution of the society so affected (Lofchie 1975). Contextualizing the problem of food shortage in terms of the socio-economic arrangements of social systems has lent a historic dimension. The character and dynamics of hunger and famine are not static and immutable.

To view the food crisis in the Sahel in this light is to highlight the failure of much foreign aid and assistance in the reconstruction and recovery of the desert edge in the postfamine period. For, in spite of the vast proliferation of institutes, research organizations, and yet more bureaucracies; of multivolume studies by the United States Agency for International Development (USAID) on 'long term development strategies'; and even a pledge by Kissinger 'to eradicate hunger', Sahelian peasants remain hyper-vulnerable and chronically poor. During the summer of 1978, the French Minister of Cooperation, M. Galley, claimed, in a press release, that another massive food crisis was imminent and that the Sahelian states required immediate food assistance to the tune of 750 000 tons (*Times*, July 12, 1978). All is not well it seems.

The geographical focus here is rural Hausaland in northern Nigeria, though much of the material presented pertains to Katsina Emirate, where fieldwork was conducted during 1976–8. The northern states of Nigeria consist largely of open, rolling Sudan Savanna fringing the desert's edge and subject to a short, intense, and highly variable wet season. In a somewhat modest fashion, I will address some of the salient aspects of the changing character of food production systems and famine in Hausaland since the 19th century. Much of the discussion will concern what could be referred to generally as social relations of food production systems and their relation to environmental uncertainty, particularly drought.[1] What I have to say needs to be grounded in Nigeria's current food crisis, reflected starkly in the sharp climb in staple foodstuff imports between 1974 ($42 million) and the first three-quarters of 1976 ($412 million). As is apparently the case in other West African states, a country with enormous economic potential and 70% of its population engaged in the agricultural sector can no longer feed itself.

Risk, hunger, and famine during the 19th century

By the end of the 18th century, Hausaland consisted of a largely Islamized population in terms of its norms and values, whose rulers were also Muslim but whose legitimacy as a dynasty was based on an ancient pre-Islamic *iskoki* belief system. From the social and political tensions of the late 18th century emerged the Holy War (*Jihād*) of 1804, led by an organized Islamic body – the *jama'a* – and committed to the overthrow of the old *sarauta* system. The *Jihād* heralded a new form of political organization, the emirate system, and a larger unified polity, the Sokoto Caliphate, which welded together 30 emirates covering 150 000 sq. miles. The caliphate survived for almost 100 years, from the accession of Usman dan Fodio as Amir-al-muminin at Gudu in 1804, to the death of Sultan Ahmadu at the hands of the British colonial forces at Burmi in 1903.

The basic unit of production in the 19th century was the household (*gida*), perhaps embracing sons, clients, and slaves in a *gandu* structure in which the householder (*maigida*) organized production and distribution and paid taxation[2] (Hill 1977). Households were usually subsumed in communities (*garuwa*) controlled through the agency of a village headman. A proportion of the peasant surplus was expropriated by a ruling class of officeholders (*masu sarauta*) in the form of either labor, grain, or cash. The *masu sarauta* had tenure over 'fiefs' given by the emir, though they usually resided on private estates worked by slaves, clients, and sometimes hired labor; they could also demand corvée (*gayya*) from villages within their territorial jurisdiction. Slave labor, though crucial to the functioning of the large estates (*gandaye*) operated by the elite and wealthy merchants, was not a dominant characteristic of the productive system. Craft production, and petty commodity production generally, emanating from within the household structure, was conversely a widespread phenomenon throughout Hausaland. The state controlled by means of coercion, provided protection for the *talakawa* (peasantry) and travelling merchants, organized large-scale labor projects and acted as a guarantor in times of need. Within this social formation, I hope to show that the nature of food scarcity assumes a historically specific character.

Disasters generally, and famines in particular, are not novel experiences for Hausaland. In the 19th century, crises of underproduction occurred in which basic biological requirements could not be fulfilled, these are well documented in the historical chronicles. During the middle of the 18th century, moreover, a disastrous series of droughts and related epidemics, which spanned a 20-year period, struck the northern savannas from Senegal to Somalia, producing economic disarray, mass evacuation from the Sahel, ethnic and no doubt monetary redistribution, and perhaps a dissolution of the large slave estates (Lovejoy & Baier 1976). It is even suggested in the local oral tradition that famine was instrumental in the evolution of the state itself. Superimposed on the pattern of major climatic disturbances were epicycles of more frequent but localized drought and food shortage occurring perhaps in the order of once every 7 or 8 years and usually regional in character. Between the *Jihād* in 1804 and the multi-year drought in 1913–14, it appears that the savannas were relatively free of an internal disaster of the scope and magnitude associated with the famine of the 1750s.

Contrary to popular belief, drought and famine are not new phenomena to the desert edge. Rather they are *recursive* and the historical landscape of northern Nigeria is littered with references to the great famines (*babban yunwa*) of the past. The dialectic of 'feast and famine' is a recurrent motif in Hausa history and the intense hungers of the precolonial period, which often embraced vast tracts throughout the savannas, were usually named and imbued with a sort of environmental personality. The recursivity of drought and food shortage is reflected in the cognitive characteristics of both phenomena in the Hausa world view. Not only is there a complex and subtle terminology, a lexicon of sorts, associated with rainfall variability and oscillations in food supply, but drought–famine is embodied in the most significant cultural and artistic forms such as praise epithets (*kirarai*), folk tales (*tsatsunyoyi*), fables (*almara*), and anecdotes (*labarunda*). The cognitive position of famine is not unlike that of the Great Depression in the West for those who lived through it. Not all famines during the 19th century had a specifically climatic etiology, but the vast majority were associated with poor rainfall and a chronic harvest shortfall.

In the light of the recursiveness of rainfall and harvest variability, it is to be expected that rural communities were in some sense geared to environmental risk, and possessed some sort of adaptive flexibility and adjustment capability with respect to drought and oscillations in the availability of food. Take the following comment from Raynault describing 19th century Nigerian Hausaland:

> Faced with precarious natural conditions indigenous society was able to place into operation a series of practices, individual and collective, which permitted it a margin of security . . . traditional techniques of storage permitted grain to be stored for relatively long periods . . . which made possible the constitution of reserves . . . after the harvest the seed destined to be planted the next year as well as the quantity of grain necessary for the subsistence of the group during the planting season were placed by the clan head in a large granary which could not be opened until after the first rains. (Raynault 1975, p. 17)

Scott has suggested that this adaptive flexibility, the capacity to cope in a risky environment, is characteristic of peasant communities generally and that precapitalist societies were to a large degree organized around the problem of risk and the guarantee of a minimum subsistence, *a margin of security* (Scott 1976). Scott refers to this margin of security as, a 'subsistence ethic', which can be subdivided into three aspects: a general proclivity toward risk aversion in agriculture ('safety first'), a tendency toward mutual support ('the norm of reciprocity'), and an expectation of minimum state

Table 6.1 Subsistence ethic and response levels in 19th-century Hausaland.

Response level	Safety first	Subsistence security Norm of reciprocity	Moral economy
agronomic/ domestic level	agronomic risk aversion intercropping, crop mixture crop rotation, moisture preservation crop experimentation, short manuring millets, etc. exploitation of local environment (famine foods) secondary resources (dry season crafts) domestic self-help and support		
community level		interfamily insurance, risk sharing extended kin groups (*gandu*) reciprocity, gift exchange, mutual support elite redistribution to the poor storage, ritual sanction antifamine institutions (*Sarkin Noma*) patron–clientage (*barantaka*) communal work groups (*gayya*)	
regional/state level			regional and ecological interdependence between desert edge and savannas local and regional trade in foodstuffs from surfeit to deficit regions (*yan Kwarami*) role of the state (a) central granaries based on grain tythe (*zakkat*) (b) state relief and tax modification

support ('the moral economy'). As a political scientist, Scott was interested in the relationships between peasant rebellion and subsistence, yet his speculation that precapitalist communities are in some sense organized around the principle of risk and environmental uncertainty is congruent with much human ecological work on human adaptation in semi-arid biomes (see Ch. 3 for a discussion of cooperation as an adaptive mechanism). Scott's three dimensions of subsistence, in fact, correspond closely to the response levels that I have identified for 19th-century Hausaland. As a human ecologist, I was concerned with the three systemic levels – the household, the community, and the state – which could buffer or mitigate the stresses imposed by drought or food shortage. A composite picture of Scott's moral economy and my own *response levels* as they pertain to the Sokoto Caliphate is presented in Table 6.1.

Response to environmental variability As Table 6.1 illustrates, precarious environments often give rise in peasant societies to a 'subsistence ethic' predicated on a safety-first or risk aversion principle. In practice, subsistence ethic might involve a plethora of locally adapted cereal varieties, a preference for the consumable versus the marketable, or a reliance on historically established planting and intercropping strategies. Throughout much of 19th-century Hausaland, the peasant economy conformed in large measure to this rough archetype. The basic agronomic strategy consisted, as it does today, of the intercropping of sorghums and millets, each characterized by contrasting moisture nutrient and shade requirements (Ferguson 1973, Usman 1974). Nicolas *et al.*, for example, have described the basic polycultural system of 19th-century Niger as a type of ecological 'marriage', which was risk averse:

> Le marriage mil-sorgho . . . est une sorte d'assurance que pratique le cultivateur; le mil s'accommode assez bien d'un hivernage assez sec; au contraire, un hivernage très humide convient mieux au sorgho. (Nicolas *et al.* 1968, p. XXII)

The descriptions by Imam Imoru (see Mischlich 1942, Hamani 1975) and other expatriate travellers confirm the impression left by Morel, that 19th-century Hausa agronomy was rational and above all scientific:

> [Hausa farmers] have acquired the necessary precise knowledge as to the time to prepare the land for sowing, when and how to sow; how long to let the land lie fallow; what soils will suit certain crops, what varieties of crop will succeed in some localities and [not] in others . . . how to ensure rotation . . . and when to arrange with Fulani herdsmen to pasture their cattle on the land. (Morel 1911)

Such adaptive behavior was supplemented by a profound understanding of local botany, such that wild resources could be exploited when necessary. A brief glimpse at the inventory of crops prepared by Imam Imoru, including an excess of drought-resistant cultigens such as cassava, lends much support to Morel's belief in the sophistry of 19th-century Hausa agronomy. The diversity of agricultural foraging, collecting, and hunting strategies bred a sort of systemic stability in a pre-eminently fickle environment (also see Ch. 1).

At a higher level, this complementarity of millet and sorghum and the agronomic knowledge of 19th-century Hausa farmers was supported by, in most locations at least, a complex orchestration of microenvironments involving variations in spacing, moisture availability, and soil type, all of which were conjoined through complicated sequential patterns of decision-making dependent upon the onset, character, and duration of the rains (see Watts 1983).

Response to food shortage The 'subsistence ethic' was also expressed through social activities and institutions which functioned as, among other things, guarantors of a minimum food supply. Fundamental to the preservation of a measure of self-sufficiency was, of course, storage which permitted the long-term constitution of reserves sufficient to cover seed requirements and grain during the period of preharvest hunger.

The subsistence ethic was also expressed through the closure of household granaries during the postharvest period, which often corresponded with the departure of adult males on dry season migration (*cin rani*), frequently as corvée labor on state-sponsored (defense) projects. During the wet season itself, when seasonal food shortages peaked, hardship could be partially alleviated by participation in communal work parties and short distance migration, taking advantage of the variation in the onset of the rains (and hence in the timing of planting, weeding, and harvest).

Central to the subsistence ethic, however, and to the moral economy in general, was what one might call the 'logic of the gift': the reciprocal and redistributive qualities, in other words, which to a large extent bind the peasant social fabric, and by which the possibility of accumulation finds an institutional obligation to redistribute. A few examples will show how important this 19th century social institution was as a guarantor of a minimum food supply.

Marcel Mauss, in his classic work *The gift*, showed that for many noncapitalist societies the 'logic of the gift' operated simultaneously as a sort of language and a means for cementing the social fabric (Mauss 1954, see also Sahlins 1972). In this way, the gift giving, what Scott refers to as the norm of reciprocity (Scott 1976), mediates against excessive accumulation through an ideology that stresses equality through redistribution. Raynault, referring to Hausaland, points out that:

> Dans un tel système, donc, la possibilité d'accumulation de travail trouve une contre-partie institutionnelle dans l'obligation de redistribuer, au moins partiellement, les richesses produites. Il s'agit, dans une certaine mesure, d'un mécanisme de sécurité collective. (Raynault 1976, p. 288)

The gift, therefore, is substantively rational in the sense that in a society in which redistribution is emphasized, it is an investment both as a means of accruing prestige and as a security in times of dearth. To a certain extent, the *biki*[3] (Hill 1972) system in Hausaland, which exists today in modified form, especially among the poorer peasantry, operates in this way. In a similar fashion, the inflated emphasis on the role of kinship and descent grouping generally serves to reinforce the ways in which risks are diffused and collective security instituted. Among the non-Muslim Hausa (*Maguzawa*) the descent group referred to as the clan segment (*GIDA*)[4] functions precisely to this end:

> [The *GIDA*] has but one function: when the grain stores of one household are exhausted, its head may borrow grain from another GIDA household and repay that grain at harvest without interest. (Faulkingham 1971, p. 81)

At an ideological level, the redistributive ethic was reaffirmed through Muslim doctrine, which saw gift-giving as obligatory for the rich (*masu arziki*) and the office holders. At another level, other formal institutional mechanisms incumbent upon the *masu arziki* served to free resources from the rich to the *talakawa*. The *gayya* or communal work group is a case in point in which foodstuffs are released during the critical preharvest period:

> [*Gayya*] peut apparaître comme un facteur de redistribution et d'équilibre de la société en ce sens qu'au moment où les ressources alimentaires commencent à diminuer et quand la disette se fait sentir . . . (Raulin 1964)

A rather more elaborate instance is the institution of *Sarkin Noma* (lit. king of farming) among non-Muslim Hausa[5] (Nicolas 1967). The *Sarkin* 100 is elected by virtue of his capacity to produce in excess of 1000 bundles of grain. In essence, it is an attenuated variant of the North American 'potlatch' in which prestige is accrued through the ceremonial distribution of resources. The office of *Sarkin Noma* entails on the one hard a redistribution of foodstuffs through the harvest festival of *dubu* and on the other a function related to seasonality itself:

> [*Sarkin noma*] is the ultimate defense against famine: when the grain in any *gida* is exhausted the residents may obtain an interest free loan of grain from the *S. noma*'s bins, to be repaid at harvest. Finally, the *S. noma* provides seed for any who may have consumed his own supply . . . (Faulkingham 1971, p. 81)

This form of ceremonial redistribution was replicated through other mechanisms, notably *Kan Kwariya* (a sort of female equivalent to the *Sarkin Noma*), *girka* (the initiation of the *bori* adept), and *wasan kara* (intervillage exchanges by youths).

In a society predicated upon an absolute hierarchical segmentation between peasant *talakawa* and *emir sarauta*, it is hardly surprising that the upper echelons of the political system in 19th-century Hausaland were expected to act as the ultimate buffers for the village level redistributive operations. The responsibilities and obligations of the village heads (*mai gari*) were quite clear in this respect and when their capabilities were overridden, in cases of extreme seasonal hardship, the next level of the hierarchy (the district chiefs or fief holder) was activated. In Katsina Emirate, for example, the district authorities often kept grain at several centers throughout their district and frequently in villages where they may have acted as *uban daki* (patron) to a number of clients (*barori*). These graduated responses terminated with the state structure itself which, as Palmer and Smith noted, used the grain tythe (*zakkat*) for central granaries for organized redistribution during famine periods (Palmer 1911, Smith 1967; see Ch. 1 for discussions of relations between the Hausa and the Fulani).

In the ways illustrated above, responses to food shortages were graduated with respect to time and depth of commitment of resources. The early responses tended to be shallow and reversible – perhaps the sale of livestock or familial loans – and the later ones less flexible, perhaps culminating in widespread dislocation through permanent outmigration or even death. All this is not to suggest a Rousseauian precapitalist bliss, a glorified peasant life somehow optimally adapted and ultrastable. Rather, I simply wish to suggest that some institutions, mechanisms, and practices – indeed some of the most prosaic attributes of peasant society – embodied in the Sokota Caliphate, provided a measure of security and buffered households from the worst effects of variability in food supply. The security arrangements were grounded in and inseparable from the architecture and constitution of the entire social formation and were indeed instrumental in the reproduction of it.

Colonial integration and the demise of the moral economy

Colonialism in northern Nigeria was a process of incorporation in which extant precapitalist modes of production were articulated with the colonial, and ultimately, the global economy (Shenton 1983). This articulation was principally effected through the colonial triad of taxation, export commodity production, and monetization. Although the colonial state left peasant producers in control of the means of production and instituted minimal technological change, the process of incorporation did necessitate a transformation in the social relations of production. To the extent that precapitalist elements in northern Nigeria were eroded by colonial integration, the adaptive capability of Hausa communities and the margin of subsistence security accordingly changed. In the process, peasant producers – particularly the rural poor – became less capable of responding to and coping with both drought and seasonal food shortage. Traditional mechanisms and adjustments disappeared, the extension of cash cropping undermined self-sufficiency in foodstuffs, a dependence on world commodity prices (for cotton and groundnuts) amplified an already high

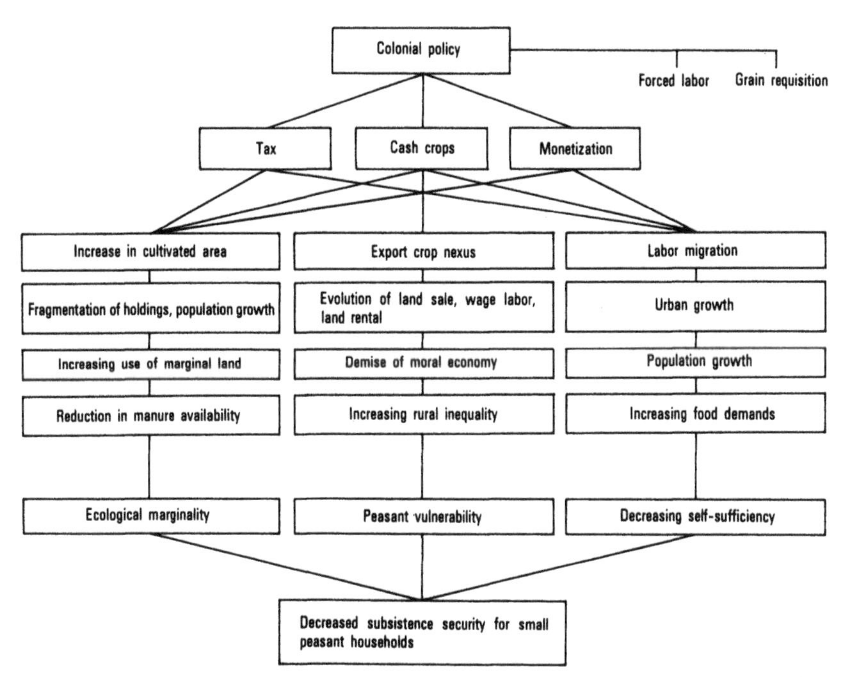

Figure 6.1 From subsistence to hunger due to colonial penetration in northern Nigeria.

tax burden, and households became increasingly vulnerable to environmental perturbations such as drought or harvest shortfalls (Fig. 6.1). This vulnerability and marginality is highlighted in four major famines which occurred during the colonial period in 1914, 1927, 1942, and 1951.

Taxation and monetization The new colonial administration sought, through taxation, to divert as much of the surplus formerly extracted by the *masu sarauta* to their own coffers. Taxes were reorganized, but for the most part they remained at the same level and in some cases revealed sharp increases to compensate the declining revenue of the *masu sarauta* (Palmer 1908). More traumatic, however, was the move to collect taxes in cash (i.e. in pounds sterling) not grain; in effect by 1910, this move not only undermined the *zakkat*-based grain reserve but:

> Il est certain que le système de taxation mis en place par l'administration coloniale a joue un rôle determinant dans le mouvement de pénétration de l'argent moderne au sein des circuits de l'économie traditionnelle. (Raynault 1977)

Furthermore, taxation had profound and direct implications for seasonal hunger itself. First, unlike the indigenous Hausa fiscal system, colonial taxes were regular, reasonably predictable, and *rigid*. The inflexibility accordingly took no account of what were the realities of Hausa life: late rains, poor harvests, seasonal hunger, and a precarious environment subject to perturbations ranging from locust invasions to epidemics and droughts. The severity of colonial taxation contrasted sharply with a precolonial *jekada* system which, though far from being innocent of extortion, made an attempt to graduate taxes according to existential circumstances (Palmer 1908, p. 69). Secondly, the *timing* of tax collection assumed colossal importance. This was especially true throughout the principal cotton-growing areas where annual taxes were gathered prior to the cotton harvest, leaving the rural cultivator little choice but the sale of grains when prices were lowest or alternatively vulnerable to the clutches of the moneylender. (It was less true where revenues from groundnuts were used to pay taxes.) And thirdly, the colonial taxation system was inseparable from the policy of encouraging the extension of commodity production and cash cropping into the countryside.

It is quite clear that the colonial taxation system in northern Hausaland, where groundnuts were the principle tax-paying crop, perhaps goes a long way to explaining the apparently 'irrational' behavior of a peasantry, who produced more groundnuts when the commodity price had actually fallen in an effort to get money to pay taxes. More generally, of course, the groundnut revolution, particularly in the close settled zones, meant a decrease in the area devoted to foodstuffs, increasing subjection to the vagaries of the world commodity market and the ever present threat of indebtedness at the hands of middlemen.

It is precisely in this way that the nature of seasonal hunger changed both in terms of its dynamics and the predicament of those who find themselves suffering from its effects. The result tended to be that seasonal hunger easily evolved into full-scale famine. Thus the *babban yunwa Gyellare* (lit. the great famine called Gyellare) of 1913–14, and indeed those which followed it with monotonous regularity up to 1974, can only be understood in terms of a social formation whose precapitalist structure has been partially transformed.[6]

Firms and the credit nexus Despite the commercial setback of the 1913–14 famine, the groundnut revolution picked up momentum and became emblematic of the subsequent expansion in the export (cash-crop) trade. Through this process of agricultural transformation from subsistence to cash-crop production and the increasingly important role that money came to play, it is hardly surprising that new forms of indebtedness arose. This was especially so in the case of the coevolution of the *'yan baranda* system and the cash-crop economy (Katsina Native Authority File 1343/4 1933). The *'yan baranda* constituted the lower order of the export crop buying hierarchy, receiving cash advances from European firms via their buying agents (*fuloti*). These sums were in turn either used to purchase produce directly or they were lent directly to the producer who pledged his crop to the agent. The interest on such loans was frequently in the order of 100% and for the producer at least was the initial step into a cyclical debt trap. The point is, of course, that through the cash advances made to them the lower order middlemen and the *sarauta* actually exploited financially weak producers due to 'normal' seasonal variations and the preharvest need for cash in particular. It is precisely in this manner that European buying agents and possibly other urban and merchant capital penetrated the Hausa countryside and it illuminates the way in which the previously subsistent family unit is drawn into an externally orientated market network. As Shenton and Freund so nicely put it:

> The most successful traders stood at the apex of a network of credit and clientage that rested on the shoulders of the 'yan baranda, the village middlemen . . . the merchant class ultimately lived in the interstices of an economy dominated by European firms. (Shenton & Freund 1978)

Commercialization and the decline in the 'spirit of the gift' Clearly, the deepening involvement by Hausa farmers with cash-crop production impinged upon the social organization of agricultural production. In the groundnut zone of Niger, this has taken the form of the dissolution of traditional estates, an escalation in land sales and the generalization of hired farm labor (Raynault 1977). Changes in the sociology of production were also coupled with the profusion of imported commodities, especially cloth, which articulated with the cycle of rapidly inflating prices for ceremonial exchanges on the one hand, and the chain of indebtedness on the other. Stresses, consequently, were imposed upon the corporateness of the rural community; the old responsibilities and obligations became less binding, collective work group largely disappeared, and the *gandu* became less embracing and hence increasingly incapable of buffering individuals during times of crisis. In the densely settled areas, for example, the extreme land shortages heralded larger food deficits and heightened vulnerability to environmental seasonality and productivity.

The extended household showed the first signs of fission and collective security had lost its original meaning:

> l'univers dit traditionnel n'a plus guère de réalité intrinsèque: les relations entre groupes humains et milieu ont subi de profonds desequilibres; les structures sociales familiales . . . s'effritent; les solidarités anciennes se relâchent; même les comportements qui semblent s'établir dans un continuité directe avec le passe – les dons par exemple – perdent progressivement leur signification originelle. (Raynault 1975, p. 13)

In short, the social nature of the subsistence system had been disrupted in the sense that the distribution and production of foodstuffs was no longer based upon the old, socially established norms of gift-giving. Reciprocity and solidarity, and hence the nature of inequality itself, had changed. Of course, Hausa society, prior to the arrival of the British, had stressed hierarchy and certain forms of inequality, but the appearance of new, colonial inequities were founded not on social structure but on changes in the social relations of production.

Given all that has been said, it is important to note that the character of agricultural production was not, and has not been, *totally* transformed by the penetration of British, merchant capital. The basic productive unit (*gandu*) remains intact, traditional tools and techniques of cultivation are still used, and the development of a landless class, deprived of its means of production, is relatively undeveloped. None the less, the pattern of changes that have occurred and the impact they have on food production are made clear, I think, by the incidence of famine and drought in Hausaland since 1900. In Katsina, for example, famine and drought occurred in 1906, 1913–14, 1920, 1927, 1931, 1942–3, 1950–1, 1954, and 1974. This uncommonly long sequence is itself astonishing, but much more to the point are the mechanics by which normal seasonal hunger assumes famine proportions. This series of shortages in fact documents the manner in which seasonal hunger is compounded by and parallels the sorts of transformations discussed by Raynault (1975, 1977).

Take, for instance, the famine, known locally as *Kwajaja*, which occurred in 1950–1. Famine conditions were precipitated, as the Katsina Native Administration was only too aware, by buying agents who, foreseeing a mediocre grain harvest, demanded that mortgaged cash crops and other debts be redeemed with grain and other subsistence foodstuffs. The peasants conversely, already burdened with high taxes and partially filled grain bins due to a poor harvest in 1949, were extraordinarily vulnerable. The Katsina Resident could, unfortunately, only lament the fact that the peasantry were essentially grainless while the foodstuff market had been effectively cornered by urban merchants. The colonial administration, still without a grain reserve policy, was therefore faced with the double bind of buying grain for relief either at inflated prices from urban wholesalers, or of incurring the high freight rates involved in purchasing the cheaper guinea corn and cassava from the southern provinces (Katsina Native Authority 1952).

The contemporary situation: the view from Kaita village

The following is an abbreviated and schematic presentation of research conducted in Kaita, a large Hausa village in Northern Katsina between 1977 and 1978. Given the exigencies of space, all that will be done here is the presentation of some salient aspects of the rural economy that bear directly on the previously analyzed problem of administratively related food shortage and scarcity in an uncertain environment.

The village of Kaita is a large nucleated settlement (pop. 2809), consisting of 494 households. It is located 15 miles northeast of Katsina town and adjacent to the River Kaita. The vast majority of household heads are farmers who produce millet and sorghum. Twenty-five percent of all farmers have access to lowland (*fadama*) farms, which are employed in the lucrative dry season irrigation (*lambu*). Upland farms have been extraordinarily scarce since the 1940s, due to population increase, and population density currently is in the order of 400 per sq. mile. As is the case in most rural Hausa communities, there is a pronounced economic inequality between households, particularly in landholdings. The poorest 20% of households are barely self-sufficient in grains during a good year because they have relatively little land to put into cultivation.

First, I want to present a synopsis of various forms of seasonality experienced by Kaita farmers to illustrate the ebb and flow of all kinds of social and economic activity.[7] (*Editor's note:* the following scenario might represent village life in any part of the Savanna–Sahel zones of West Africa.)

By the end of the dry season, the labor and energy required for fetching water, gathering food, and off-farm activities (to earn cash to purchase food) increase. Food becomes scarcer, less varied, and more expensive in the marketplace. Poorer families, who have smallholdings, or who have limited familial labor, begin to suffer. As the dry season intensifies, they have less food due to low productivity, labor supply, and capital and they are also unable to profit from lucrative nonfarm occupations. Some poor farmers migrate during this time; others undertake casual labor at low wage rates.

The rains bring crisis. Heavy and urgent agricultural demands have to be met and prompt timing of agricultural activities is essential. Intense physical labor requirements – for land preparation, planting, and weeding – come during 'the hunger period' when food supply is lowest. Food prices are now inflated and rural communication is difficult. For those with land, future food supplies depend upon the ability to work or hire labor during this crisis period. Many of the very poor are losing weight because their work output exceeds their calorie intake. In addition, the quality of the nutrient intake declines.

The wet season is also the least healthy period of the year, with maximal exposure to infections. Government health services are often restricted at this period when natural immunity is low. Births tend to peak during the wet season, making it difficult for women and small families.

Afflicted by sickness, birth and pregnancies, food shortage, poor diet, and high grain prices, the poor are usually in desperate need of cash and are driven to distress borrowing. They go to moneylenders or patrons; they may mortgage their crops (prior to harvest) or indent their labor in exchange for food. In essence, the seasonal energy crisis shifts the terms of trade against their labor and pushes them into financial dependence. Fear of future crises perpetuates this dependence and the poor are screwed down seasonally into subordinate relationships through their vulnerability and exploitative relations of production.

When the harvest comes, debts have to be repaid, along with taxes. Food is abundant, but prices are low and debts must be repaid in a buyers' market. As the dry season progresses, conditions improve, prices rise, food is available, though high priced, risk of infection declines, caloric intake rises, and diets become more varied. Those who lost weight regain it. Marriages, ceremonies, and social activity increase and there is a peak in inception rates. Gradually, the cycle commences all over again.

There are, of course, deficiencies in the scenario. None the less, it is a fair representation of the seasonal rhythms in Kaita village and serves to highlight the fact that each wet season is a crisis period, with a conjunction of trends, patterns, and activities that amplify already latent stresses (Table 6.2), whereas each dry season is, relatively speaking, an abundant period. The context, which a seasonal perspective provides, illustrates the vulnerability of marginal (poor) households *during the 'normal' or regular cycle of events*. A quantitatively small perturbation, superimposed upon the

Table 6.2 Seasonality and stress in Kaita.

Factors	Forms of stress	Dry season			Wet season			
		Early	Mid	Late	Early	Mid	Late	Harvest
						Crisis period		
diseases	cerebrospinal meningitis			–				
	malaria					—	—	–
	diarrhea					—	—	–
	guinea worm					—	—	–
	skin infections					—	—	–
	filariasis	–						
energy, food, and nutrition	agricultural energy demand	(–)			–	—	—	–
	agricultural energy demand by men		+		–	—		–
	food stocks	+	+		–	—	—	+
	prices for food purchase	+	+		–	—	—	+
	food quality	+	+			—	—	+
	body weight/energy balance	+	+		–	—	—	–/+
economic	debt and repayment factors			(–)	–	—	—	–
social and demographic	child care	+	+		–	—	—	–
	deaths	–	+	+		—	—	–
	neonatal deaths as % of births					—	—	
	conceptions		H	H				
	births						**H**	H

+ = a positive condition or effect; – = a negative condition or effect; H = high.
Source: adapted from Chambers (1978).

seasonal rhythm, whether it be drought, locust invasion, or a sharp decline in food availability, has a correspondingly large repercussion on farmers' wellbeing. The catalytic role of drought, or indeed any environmental change, in the process of deepening economic polarization and even pauperization of the rural poor is now clarified.

The poor, shackled by their poverty, find themselves in what Waddington calls 'lock-ins' (Waddington 1977); that is, the poor are tightly constrained by their capital shortage and smallholdings and are placed in dependent relationships with the wealthier householders. Indeed, it is the poor who are powerless and must dispose of minimal assets at undervalued prices. Response to stress thus appears as an ineluctable, one-way street, in which the poor reluctantly dispose of their Lilliputian resources. It is, as Chambers calls it, a 'ratchet effect' in the process of impoverishment (Chambers 1978). But when Richards suggests that hunger is a question of distribution and administration of foodstuffs (Richards 1975), he is only partly correct. Hunger is also firmly embedded in the peculiar conjunction of seasonality and the *productive* system itself. This is not to suggest that seasonality or variability in these seasonal rhythms *causes* poverty or *generates* wealth. Rather it is the failure of the 'modified' society in question to cope with their seasonal and environmental effects. Seasonality in Nigerian Hausaland provides part of the context within which social processes operate and, perhaps, the occasion for intensifying rural inequality, thereby amplifying seasonal stresses.

An analysis The relationship between seasonal rhythms and poverty is critical in a context such as the one presented here, where interhousehold income differentials and access to lucrative trades and occupations are so pronounced. In Kaita, as elsewhere in Hausaland, the disparity between the yields per unit area on rich and poor farm holdings is marked, due largely to the differential application of organic and artificial fertilizers. For the low income households with minimal livestock resources, the disappearance of the *Bororo* (nomadic) Fulani and the prevailing poor, aeolian soils in the district contribute to the already disappointing farm productivity. In addition, it is the poor who are frequently compelled to sell manure to supplement the household finances.

I shall identify three developments that seem to both highlight and amplify the vulnerability of some farming families to environmental fluctuations and variability in food supply. They are by no means exhaustive, but rather they make clear some of the relations between political economy and hunger in the Hausa countryside.

INCIPIENT LANDLORDISM One of the principal ways in which household heads attempt to alleviate the attenuating circumstances of preharvest scarcity, especially in view of the historical decline in rurally based crafts, is through dry season irrigation (*lambu*). The pattern of ownership of *fadama* land is, however, markedly unequal (see Table 6.3). Since the lowland,

Table 6.3 Pattern of *fadama* ownership.

total land area	282 174 sq. yards
total number of farms	121
area occupied by the largest 24 (22%) farms	116 546 sq. yards (41%)
area occupied by the smallest 65 (60%) farms	83 811 sq. yards (29%)

fadama area is spatially limited and almost entirely cultivated at present, a system of land rental has emerged. *Lambu* (dry season irrigation) is widely practiced and suitable land is in great demand. It swells enormously when the millet harvest has been poor. Almost 40% of the plots were rented in 1978. This phenomenon is not simply indicative of rural differentiation along lines of relative wealth, but rather expresses a new kind of relationship. Patronage and kinship naturally affect landlord–tenant relationships but in many cases they are relatively impersonal and transitory. Interestingly, the payment of rent (sometimes referred to as *kantahuda*, in which the tenant pays the landlord a proportion of the *lambu* produce) is frequently denied and is conceived as a 'borrowing' agreement (*aro*). This ideological obfuscation has been described by Feldman in Tanzania:

> [reference to borrowing] is characteristic of both the way renting is rationalised by landlords and of the system of rental. This is on an annual basis and the casual expectation of the tenant with regard to security . . . but both the reasons for renting and for leasing out of land are quite different from the borrowing of land . . . in former times. (Feldman 1975)

It is precisely because these sorts of relationships are seen as *aro* that potential class conflict or domestic strife is contained. Not only are the poorer households denied access to this rental system, since a prospective landlord views a poor household with financial scepticism, but labor constraints characteristic of these familial units effectively prevent them from engaging in dry season agriculture, even when land is available. Furthermore, for ideological reasons, when poor households can hire or rent lowland farms the payment is frequently in *labor time* during the following wet season which deflects energies away from the crucial farming activities (i.e. the timing of planting and weeding) on their own holdings.

RURAL INFLATION Since the Udoji Commission and the oil boom of the early 1970s, northern Nigeria has been in the grips of a staggering rate of inflation, which has spilled into the countryside with profound implications for the poor. Not only has there been a secular decline in the terms of trade with respect to commodities that have become essential parts of the household budgets in rural Hausaland, but more seriously ceremonial expenses have rapidly outpaced the capacity of the poor to pay. Marriage is a case in point because the formalized exchange of gifts between bride and groom – a large part of which now consists of imported wares – is enormous in relation to the purchasing power of the poor. For example, an 'average' wedding expenditure in Kaita village would currently involve an outlay of N400 ($600), which is far in excess of the financial capacity of the

Table 6.4 Marriage expenses: Kaita village, 1977–8.

	N	N
alama: sign of affection	10.00	
gaisuwa and greetings gifts	5.00	
jin magana: discussion with parents	10.00	
Na gane ina so: "I desire her"	50.00	
a small *lefe*	20.00	
tashin sallah: presents for religious festivals	20.00	
baiko: confirmation ceremony, when gifts of mats, jugs, utensils, etc. are given	50.00	
lefe: clothes for the bride	100.00	
dauren aure: marriage ceremony and food expenditure	60.00	
tarewa: (partnership) gifts for the bride's arrival at the groom's house	30.00	
walima (marriage feast): this would include millet, sorghum, and meat, gifts for wife's friends, and hairdressing	30.00	
subtotal[a]		390.00
parents will have to match the *lefe* price of the groom's parents to provide property for the new wife	100.00	
kayan gara: money given to the groom and his parents at the feast plus property carried by cavalcade of women accompanying the new wife to her husband's house	150.00	
subtotal[b]		250.00
TOTAL		640.00

[a]Marriage expenses of the groom met by groom's parents for the first wife (assuming he is living and farming with his family).
[b]Marriage expenses of the bride's parents for the marriage of their daughter.

poor (Table 6.4). The net result is that sons from poorer families are increasingly incapable of obtaining a wife or the householder is pushed into debt or farm sale to cover marriage expenditures. Since expansion and consolidation of the household through marriage is one of the ways in which the poor attempt to acquire some form of economic security, rural inflation has had a profoundly deleterious effect on their welfare.

MERCHANT CAPITAL AND THE SOCIAL RELATION OF TRADE The third point focuses on the penetration of British merchant capital, the rural debt it caused, and the structure of the trade imposed by the export companies (Fig. 6.2).

Figure 6.3 indicates that poor Hausa farmers often find themselves in the double bind of having to sell cheap and buy dear. To obtain cash for taxes, ceremonies, and domestic expenses, many poorer households sell foodstuffs

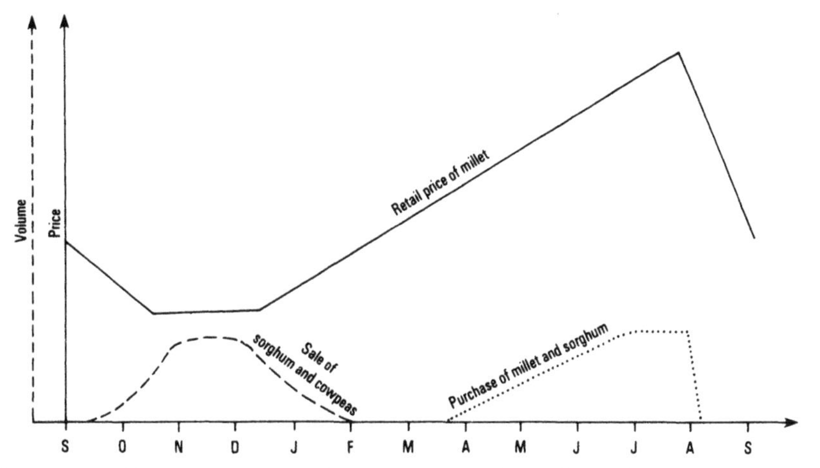

Figure 6.2 Pattern of seasonal grain sale and purchase for a poor peasant household.

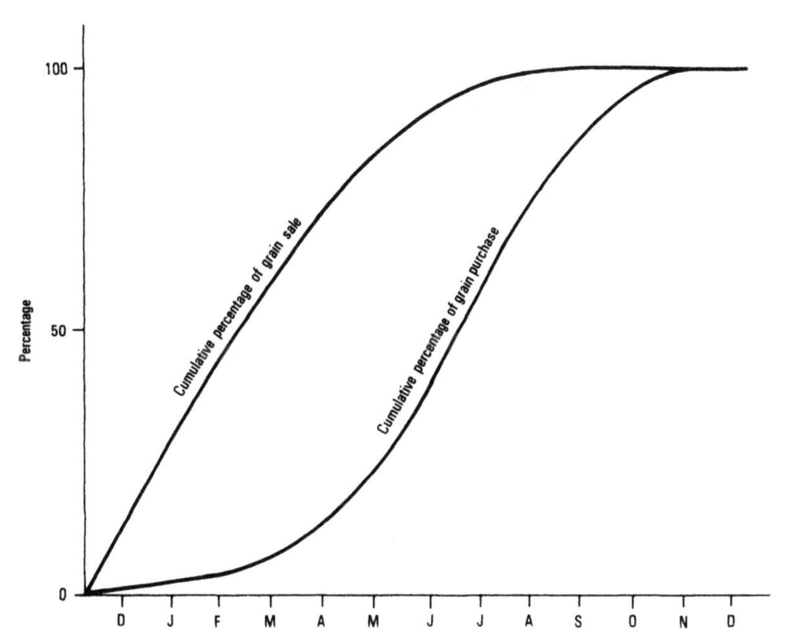

Figure 6.3 Cumulative percentages of grain purchase and sale for small peasant households.

in September–December during a general glut at low prices. Grain traders are, of course, particularly active during this period, buying for a subsequent price rise. In the event of a mediocre harvest, those households which are short of grain are compelled to purchase between May and August at inflated price levels. Many resort to borrowing either money or grain during this period (*bashida ruwa*)[8] at highly usurious rates of interest, usually from rich grain traders. These merchants are part of an enormous network of commercial clientage which bifurcates in extended chains being financed from urban-based merchant capital. This so-called 'antediluvian' capital supports the merchant nexus and finances the credit system at the local level. Whereas the poor enter into debt at usurious rates of interest, the middle peasantry, conversely, tend to exhibit bimodal selling patterns:

> they contribute to the post harvest glut because of heavy social pressures for ceremonials . . . However they tend to sell relatively less and . . . borrow more than poor farmers. During the [wet] season grain will be sold to [finance] wage labour. (Harriss 1978, p. 4)

The rich farmers withhold grain for a mid-season price rise (*ciko*), which they then sell to buy wage labor and to invest in cattle which yield enormous speculative profits. The same large farmers often buy and sell grain which, in view of the volatile nature of local markets, can generate high rates of return on capital over a short time period:

> the social relations of trade thus force most farmers to sell grain when prices are low or to go into debt to be repaid at harvest at high interest rates. Through this market structure a hierarchical trading system financed by urban merchants and hereditary nobles and tied by patron–client dependency relations accumulates large financial and physical surpluses through its control of production. (Harriss 1978, p. 5)

Raynault was therefore quite correct in his supposition that cereal commerce did not signal the existence of a surplus, for everything pointed to its origin in the economic vulnerability of some households and its operation to accentuate their weakness and dependence, fueled by their need for money.

Conclusion

In contrast to much research which emphasizes certain leveling mechanisms at work in Hausa communities, a seasonal perspective on rural economy highlights the growing polarity between economic groups and the genesis of qualitatively different social relationships. This process of peasant differentiation is related to the hiring of labor, usury interest, and what is sometimes called antediluvian capital. These exploitative relations are quite frequently disguised at the village level, as Wood has observed, through verbal agreements, informal arrangements and, more pervasively, through a 'hegemonic, egalitarian Muslim ideology' (Wood 1978). It is precisely because of the social relations of production that some households are increasingly incapable of responding to food shortages and, in view of the atrophy of many of the traditional buffering mechanisms, a drought may act as a catalyst by which economic inequalities are amplified and the slide into impoverishment is hastened.

Hill is nevertheless correct when she observes that there are some village responsibilities for the provision of social security (Hill 1972). There is a widely held ethic among rural communities that the poor should be assisted in times of need, and that the traditional grain tythe, the *zakkat*, does constitute a major component of grain circulation. But it would be a mistake to glorify this gratuitous dimension of village life and inflate the role of traditional leveling mechanisms. It is rather that few people actually starve in Hausa villages but there are many who are chronically and debilitatingly poor and shackled by their poverty. The self-perpetuating poverty trap makes it almost impossible for those born into desperately poor families to improve their lot. The fact that rich mens' sons can slide into poverty, or may inherit a small slice of the familial pie, does not alter this fact. One should not let the collective responsibilities, such as they are, obscure the predicament of the rural poor. Indeed, if Hill's work in the 'stagnating, impoverished, miserable circumstances' of the Kano close settled zone is any precursor of what is to come (Hill 1977), the future of rural northern Nigeria may not be as sanguine as some observers would have us believe.

Haswell's description, which follows, shows that the situation in Nigeria is not unusual in other parts of West Africa:

> This system has a most sinister aspect, however, when it is applied to a largely monetised economy, so long as the community continues to accept as equally legitimate a 100 per cent repayment over the short term on a whole range of petty trader and moneylender transactions which merely enslave it in a perpetual round of debt; for as consumer goods become available in local stores, which producers deriving their income from a low-productivity agriculture have not the means to procure but are encouraged to purchase, persistent poverty results in production forgone (through consequential inefficiency and ill-health) which eventually becomes a cost to society. (Haswell 1975)

Notes

1 This is part of a much larger study (see Watts 1980).
2 This section draws heavily on Shenton and Freund (1978).
3 Polly Hill says that *biki* is 'a very important socio-economic institution linking virtually the entire population' (Hill 1972, 1977).
4 *GIDA* here refers to the clan segment, distinct from *gida*, the household. Patrilineages and clan segments are peculiar to non-Muslim Hausa.
5 On the institution of *Sarkin Noma* in northern Nigeria, Reuke writes:

> In the traditional political system of the Maguzawa there were several officials . . . who are today of little significance. The most important of these officials and his associates were: *Sarkin Noma*, Lord of the Harvest, *Sarkin Hu'da*, Lord of the Furrows, *Sarkin Dawa*, Lord of the Bush, and *Sarkin Arna*, Lord of the Heathens. Their jurisdiction is the kasa or village district . . . [*Sarkin Noma*] generally possesses a larger piece of land than the other *masugida* and as a result controls a considerable harvest surplus. If the grain runs out in a *gida* before the new harvest, the *maigida* takes himself to the *Sarkin Noma* and asks for help. Either grain would be sold to him or loaned until the next harvest. Interest would not be paid on its return. (Reuke 1969).

6 Unfortunately, little information is available in the archives, since the colonial administration was for the most part ignorant of the food situation. This is reflected in the paucity of references to famine in the Annual Reports and by the staggering remark of the Kano Resident on the 1908 famine: 'Yes, the mortality was considerable [but] we had no remedy at the time and therefore as little was said about it as possible.' (Katsina Native Authority 1909). However, take for instance the following extract from the Zaria Annual Report of 1918 on the food shortage of that year: 'The food shortage is said to be due to two causes, the increased cultivation of cotton and groundnuts to the detriment of corn, and the ever-increasing non-farming communities on the Plateau . . .' (Katsina Native Authority 1919).
7 This is a modification of a position paper given at the Conference on Seasonality and Rural Poverty, Institute of Development Studies, Sussex, July 1978 (see Chambers 1978).
8 There are several types of loan arrangements. *Falle* generally refers to grain borrowed and to be repaid in kind at harvest, usually at the rate of two returned for one borrowed. *Bashi da ruwa* pertains to any loan on which interest (*riba*) is paid.

References

Baker, R. 1977. *The Sahel: an information crisis.* Vol. 1: *Disasters,* 13–22.

Chambers, R. 1978. *Seasonal dimensions to rural poverty*, p. 11. Paper presented to the Conference on Seasonality and Rural Poverty, Institute of Development Studies, University of Sussex. Mimeo.

Faulkingham, R. 1971. *Political support in a Hausa village, Niger*, p. 81. PhD. thesis. Michigan State University.

Feldman, R. 1975. Rural social differentiation in Tanzania. In *Beyond the sociology of development: economy and society in Africa and Latin America,* I. Oxaal *et al.* (eds), 181. London: Routledge & Kegan Paul.

Ferguson, D. 1973. *Imam Imoru, being a description of Hausaland during the 19th century.* PhD thesis. University of California, Los Angeles.

Hamani, D. 1975. *L'adar precolonial (Republique du Niger)*. Études Nigeriennes, no. 38. Niamey: Institute de Recherche en Sciences Humaines.
Harriss, B. 1978. *Going against the grain*. Paper presented to the International Geographical Union Conference on Marketing and Exchange, Zaria, Nigeria, July 1978.
Haswell, M. 1975. *The nature of poverty*, 209. New York: St Martin's Press.
Hill, P. 1972. *Rural Hausa: a village and a setting*, 211. London: Cambridge University Press.
Hill, P. 1977. *Population, prosperity and poverty; rural Kano 1900–1970*. London: Cambridge University Press.

Kates, R. *et al*. 1977. *Population, society and desertification*. A/CONF/74/8. Nairobi: United Nations Desertification Conference.
Katsina Native Authority 1909. *National Archives of Kaduna*, SNP 10, 472.
Katsina Native Authority 1919. *National Archives of Kaduna*, SNP 10, 95.
Katsina Native Authority 1933. File 1343/4, now located in Katsina museum.
Katsina Native Authority 1952. *Famine relief and corn reserves*. National Archives of Kaduna, 2/3, 403.

Lofchie, M. 1975. Political and economic origins of African hunger. *J. Mod. Afr. Stud.* **13**, 55–67.
Lovejoy, P. and S. Baier 1976. The desert side economy of the central Sudan. In *The politics of natural disaster*, M. Glantz (ed.), 145–75. New York: Praeger.

Mauss, M. 1954. *The gift*. Glencoe, Illinois: Free Press.
Mischlich, A. 1942. *Uber die Kulturen im mittel-Sudan*. Berlin: Andrews & Steiner.
Morel, E. 1911. *Nigeria: its peoples and its problems*, 234. London: Cass.

Nicolas, G. 1967. Une forme attenuée du potlatch en pays hausa: le dubu. *Economies et Sociétés* **5**, 151–214.
Nicolas, G., H. Magaji and M. Mouche 1968. *Étude socio-économique de deux villages Hausa: enquête en vu d'un aménagement hydro-agricole, vallée de Maradi, Niger*, XXII. Études Nigeriennes, no. 22. Paris: Centre National de la Recherche Scientifique.

Palmer, H. R. 1908. *Changes in taxation in Katsina Division*. Katsina Province 1289.
Palmer, H. R. 1911. *Kano Annual Report 1910*. National Archives of Kaduna SNP10.

Raulin, H. 1964. *Techniques et bases socio-économiques des sociétés rurales nigeriennes*. Études Nigeriennes, no. 12. Paris: Centre National de la Recherche Scientifique.
Raynault, C. 1975. Le cas de la région de Maradi (Niger). In *Sécheresse et famine du Sahel*, J. Copans (ed.), vol. 2, 5–42. Paris: Maspero.
Raynault, C. 1976. Transformation du système de production et inégalité économique: le cas d'un village haoussa (Niger). *Can. J. Afr. Stud.* **10**, 279–306.
Raynault, C. 1977. Circulation monétaire et évolution des structures socio-économiques chez les haoussas du Niger. *Africa* **47**, 160–71.
Raynault, C. *Transformation du système du production et inégalité économique*, 305.
Reuke, L. 1969. *Die Maguzawa in Nordnigeria* (trans. W. Freund), p. 25. Bertelsmann Universitatsverlag, Bielefeld.
Richards, P. 1975. Editorial. *Afr. Envir.* **1**, 4.

Sahlins, M. 1972. *Stone age economics*, 149–276. Chicago: Aldine.
Scott, J. 1976. *The moral economy of the peasant*. New Haven: Yale University Press.
Shenton, R. 1982. *Studies in the development of capitalism in northern Nigeria*. PhD thesis. University of Toronto.

Shenton, R. and W. Freund 1978. The incorporation of northern Nigeria in the world capitalist economy. *Rev. Afr. Polit. Econ.* no. 13, 8–20.

Shenton, R. and M. Watts 1979. Capitalism and hunger in northern Nigeria. *Rev. Afr. Polit. Econ.* nos. 15–16, 53–62.

Smith, M. G. 1967. A Hausa kingdom: Maradi under Dan Baskore. In *West African kingdoms in the nineteenth century*, D. Forde and P. Kaberry (eds), 99–122. London: Oxford University Press.

Time 1977. The creeping deserts, September 12, 20–3. New York.

Times 1978. July 12. London.

Usman, Y. B. 1974. *The transformation of Katsina, c. 1796–1905.* PhD thesis. Ahmadu Bello University, Zaria.

Waddington, C. 1977. *Tools for thought.* St Albans: Paladin.

Watts, M. 1980. *A silent revolution: the changing character of food production in northern Nigeria.* PhD thesis. University of Michigan.

Watts, M. 1983. *Silent violence: food famine and peasantry in northern Nigeria.* Berkeley: University of California.

Watts, M. and R. Shenton 1983. The evolution of hunger in Nigeria. *Rev. Afr. Polit. Econ.*

Wood, G. 1978. Class formation and antediluvian capital in Bangladesh. *Institute for Development Studies Bull.* **9**, 42.

7 *Changing land-use patterns in the* fadamas *of northern Nigeria*

BERYL TURNER

In the tropical rainforest of West Africa, floodplains and marshy valley bottoms are generally avoided for agriculture. Swamp rice is the only crop that can be grown in them without major drainage operations and since rainfall in the region is adequate, or even excessive for most crops all the year round, the floodplains and valley bottoms are avoided. At the opposite extreme, on the desert margins, the valley bottoms or enclosed depressions are the only areas in which cultivation is possible at any time of the year. The Savanna areas lie between these two extremes. Rainfed agriculture is possible for part of the year, but as the semi-arid areas are approached, the choice of crops that can be grown decreases and the risk of crop failure increases. The floodplains, valley bottoms, and enclosed depressions, which can be cultivated for most or all of the year, are therefore highly regarded in the Savanna as valuable assets.

Fadamas may be defined as areas that are seasonally waterlogged or flooded.[1] They include low-lying areas adjacent to streams, and depressions without stream channels, but do not include permanently waterlogged swamps. In the extensive plateau areas covering most of the Savanna zones of Africa, the normal valley form is a gently sloping concave depression, mainly grass covered, in contrast to the woodland or shrub vegetation on the interfluves, but sometimes containing riparian woodland along the stream course. The sources of most streams are extensive seepage hollows, or *fadamas* without stream channels, which may be over a kilometer wide. These usually narrow at the point where the stream channel appears, but unless there is a steepening of the downstream gradient, or a rocky or eroded stretch, a streamside *fadama* may continue and eventually widen into a floodplain *fadama*.

All land that is not *fadama* or swamp is referred to as upland. A distinction can also be made between seasonal and perennial *fadamas*. Seasonal *fadamas* are those in which the water table falls to below 3 m (10 ft) – the maximum rooting depth of crops – before the end of the dry season. Perennial *fadamas* are those in which the water table remains above this level, and which can therefore be cultivated throughout the year.

Fadama agriculture is complementary to upland agriculture and lessens the impact of food shortages both by extending the wet season growing period into the dry season and by increasing the variety of crops that can be grown. Some crops, such as sugar cane, which has a 12 month growing season, and swamp rice, which requires flooding, can only be grown in the *fadama* or in large-scale upland irrigation schemes. Other crops such as tomatoes and many vegetables only grow well in the dry season because of the prevalence of mildew and other diseases and pests in the wet season. The *fadamas* also provide essential grazing for cattle and other livestock, both in the dry season when the upland grasses die and, in the intensively cultivated areas in the wet season, when upland areas are occupied by crops. In the wet season the *fadama* grasses are cut and fed to the animals which are kept tethered to avoid damage to the growing crops.

During years of below average rainfall, the *fadamas* play an additional role. If the start of the rains is late, additional crops can be planted in the *fadamas*, which are often neglected in the wet season. Similarly, at the end of the rainy season, if it appears that the upland crops will fail to reach maturity, it is often possible to plant a rapidly maturing variety of maize in the *fadamas*.

The *fadamas* have many other uses in addition to agriculture and grazing. They are used for fishing, hunting of small mammals, and gathering of wild plants for food, for medicinal purposes, and for making mats, fences, and baskets. The *fadama* soils also provide raw materials for building purposes. They have a higher clay content than the upland soils, and can often be used for brick making. Sand is sometimes found in the actual stream channels and is also extracted in the dry season for building purposes.

The *fadamas* are therefore of great importance in the savanna regions. They comprise an extensive natural resource covering about 10% of the land surface. This chapter considers three aspects of the land use of the *fadamas*. The first part consists of a detailed study of the *fadamas* in part of central northern Nigeria. The second part deals with the impact of technological development on *fadama* cultivation, and the third part considers the potential for further development.

Case study of the area between Kano and Kaduna

Description of the area A detailed study has been made (Turner 1977) of an area of 19 000 km^2 in northern Nigeria, between latitudes 10°30′N and 12°N and longitudes 7°30′E and 8°30′E (Fig. 7.1). The area lies partly within the Northern Guinea Savanna zone and partly in the Sudan Savanna zone, the boundary between the two coinciding with the major divide between the Chad and Niger drainage. The rainfall in the area increases gradually southwestwards from 885 mm (35 in) in Kano to 1288 mm

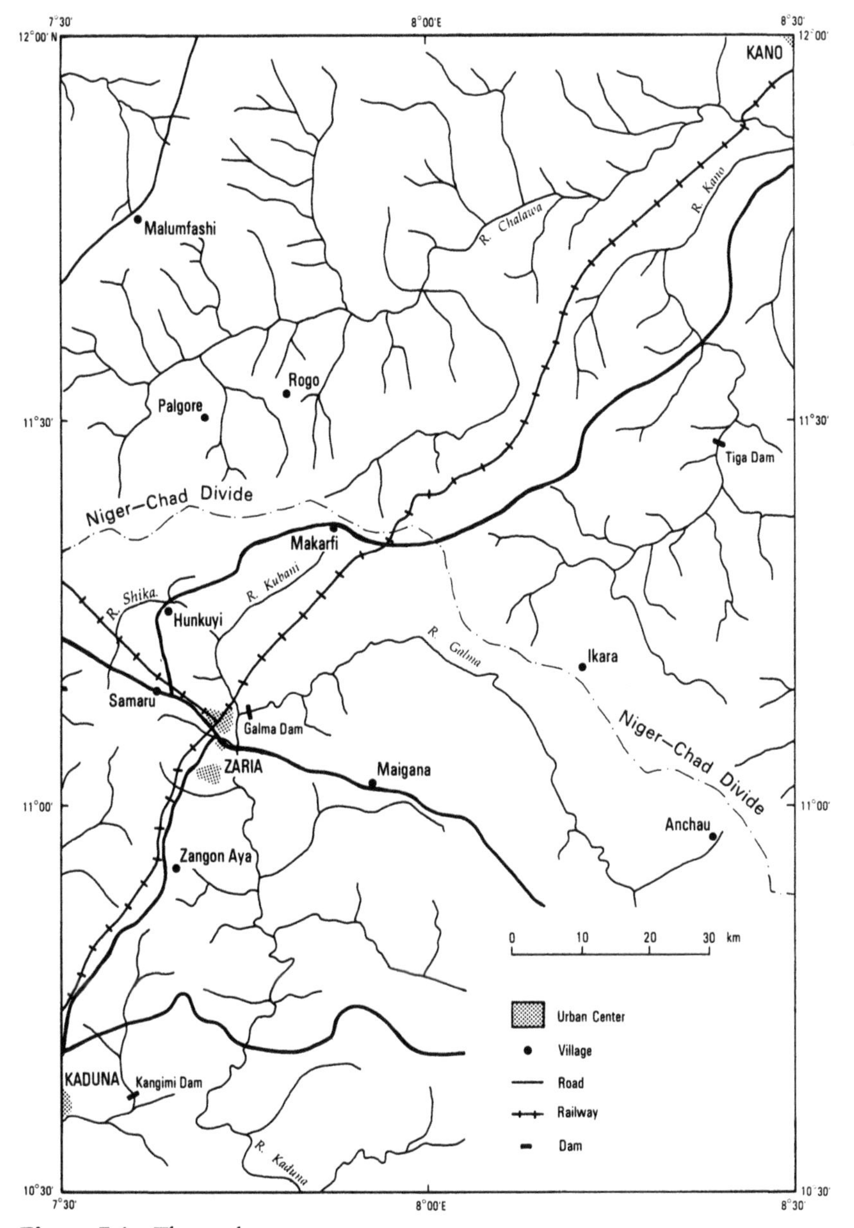

Figure 7.1 The study area.

(50 in) in Kaduna, with the length of the dry season decreasing in the same direction from 7 to 5 months (Fig. 7.1).

The area is mainly underlain by Precambrian rock with small areas of Jurassic rock in the east. The latter form hills but the remainder of the study area is part of an extensive plateau with very gentle slopes at heights between 700 m (2300 ft) along the central Niger–Chad divide, and 488 m (1600 ft) in the northeast. The soils developed over these rocks belong to two main groups: brown and reddish-brown soils of arid and semi-arid regions occur in the northeast, whereas ferruginous tropical soils cover the rest of the area. These soils are influenced not only by the underlying rocks (and the climate) but by the presence of a covering of aeolian drift, derived during a more arid period from the southern part of the Sahara to the northeast. This drift varies in thickness, usually from 0.6 to 1.8 m (2–6 ft), and in texture from fine sand to silt, obscuring the underlying rocks in the northeast and becoming both thinner and finer grained towards the southwest. Both groups of soils have a sandy or sandy-loam texture with pH values about 5–6, and low organic matter content (<1%). In the first group base saturation is higher, leaching is less, and the organic matter better distributed throughout the profile, than in the second group.

The *fadama* soils, which are derived partly from the upland soils, are very variable in composition, but are generally deeper, over 150 cm (60 in), have a higher clay content than the upland soils (16–37% compared with 10–28%), and are richer in nutrients and organic matter content (mean 1.5% compared with 0.5%). Soils in the small *fadamas* are usually loam or sandy loam at the surface, changing to clay or clay loam below 100 cm (40 in). On the larger floodplains the soils are even more variable, generally with sandy soil on the levees and clay soils in the backswamp areas, but there are sudden changes in the soils, both horizontally and vertically, resulting also in variations in water content. These sudden changes pose problems for farmers in finding crops best adjusted to each soil type.

Another feature of the study area is that it is crossed centrally by the major divide between streams flowing northeast to the Kano and Chalawa Rivers and then to the inland drainage basin of Lake Chad, and by those streams flowing south to the Kaduna River and eventually to the Niger and the Atlantic. This divide is an extremely important boundary, separating areas with very different geomorphological characteristics (Turner 1977, pp. 191–6). More importantly, there is a great contrast in the amount of *fadama* north and south of this divide, with only 5% of the area north, and 15% of the area south of the divide occupied by *fadama*.

The people living in the northern part of the study area are mainly Muslim Hausa or Fulani, the latter being traditionally migratory, cattle-owning people. The southern part of the area, around Kaduna, is inhabited by a number of minority groups, which suffered from slave-raiding by the powerful emirates further north. This southern area remains sparsely populated, with densities of about 5–10 per km^2 (12–25 per sq. mile), whereas the area around Kano has one of the highest densities of rural population in Africa, exceeding 232 per km^2 (600 per sq. mile) in a zone extending about 24 km (15 miles) from the city. There are three urban centers in the study area: Kano, Zaria, and Kaduna. The first two of these are ancient walled Hausa cities with modern towns now growing outside the walls containing large numbers of migrants from other parts of Nigeria as well as from other countries. Kaduna is a more recent town, built as the capital of northern Nigeria early this century.

Fadama *cultivation*

The areas around Kano, Zaria, and the larger villages are cultivated annually, the main crops being sorghum, millet, and beans with some maize and cassava, and with cotton and groundnuts as the main cash crops. Further south bush fallowing is practiced and there are areas of uncultivated woodland Savanna remaining.

Early fadama *cultivation* Records of cultivation of the *fadamas* in northern Nigeria date back several centuries. According to Leo Africanus, rice has been cultivated in northern Nigeria since the 16th century (Africanus 1600), and Clapperton reported in 1825 that in Zaria 'date trees, palm oil trees, papaw, melons, plantains, Indian corn, millet, *dourra* [sorghum?], rice, yams, sweet potatoes, etc. are in abundance, particularly rice. They say they raise more and better rice than all the rest of the Hausa put together' (see Hogben & Kirk-Green 1966). The original red-grained rice (*Oryza glabberima*) was grown before the Second World War as a luxury crop, eaten only by chiefs. A native beer was also made from rice and honey, but very few people could afford it and the area under rice must have been very small at that time. Expansion of rice-growing was reported by Dalziel in 1937 and demand for food by the Nigerian army during the Second World War led to greater expansion, mainly of *Oryza sativa* (Extension and Research Liaison Service 1971), which is hardier and has largely replaced *O. glabberima* except in areas susceptible to drought. Expansion continued until in the mid-1970s Nigeria was approaching self-sufficiency in rice, but demand is still rising.[2] Wheat has also been grown in *fadamas* for several centuries in small quantities as a luxury product (Andrews 1968), but rapid expansion of wheat since 1959 has been due to its production on government irrigation schemes.

Irrigation on *fadamas* has been practiced for at least 130 years (Johnston 1967, Last 1967). The earliest method of irrigation was by drawing water from wells or streams in buckets, and pouring it on the land. Shadoofs are now common throughout the Savanna of northern Nigeria and are probably the result of Arab influence reaching the study area by means of the ancient trans-Sahara trade routes. It is, however, intriguing that other methods of irrigation, such as the use of animals to draw water from wells, or the Persian wheel, are not found in northern Nigeria, although just across the border in Niger animals are frequently used for drawing water.

The ancient Hausa cities such as Kano and Zaria include considerable areas of farmland, both upland and *fadama*, within their walls, originally to maintain food supplies during times of siege and to give protection to inhabitants of the surrounding countryside (Kirk-Green 1962). The *fadama* lands were of greatest importance during the dry season when most raids took place, to provide both crops and grazing. It is not clear how much *fadama* cultivation took place outside the cities during the pre-British period, but after the Pax Britannica in 1903, there was a spreading out of the population from the walled cities into the surrounding countryside. *Fadama* cultivation increased, and with it, the use of shadoofs.

Present cultivation The *fadama* fields are usually very small (average size 0.2 ha (0.6 acres)) and are enclosed by live hedges, thorny branches, or fences made of sorghum or millet stalks to protect the crops from donkeys and other untethered animals. A very wide variety of crops is grown in the *fadamas*, some of which can only be grown in the *fadamas*, whereas others are grown as both *fadama* and upland crops at different times of the year. Some crops are grown for home consumption, but a greater proportion of *fadama* crops than upland crops are grown for sale, mainly in the urban markets.

Crop distribution Within the region there are considerable variations in the cropping patterns in the *fadamas*. A distinct zoning of the crops is apparent in many *fadamas*. Rice is usually grown in the wettest areas, with sugar cane or vegetables on slightly higher areas to avoid waterlogging and flooding, and cassava is usually grown along the margins. In *fadamas* where flooding is unlikely, sugar cane may be grown in the central areas where the roots can reach water throughout the year. In some areas where the central part of a *fadama* is subject to flooding, it is left uncultivated and used for grazing, with crops grown only along the margins. Planting of some crops such as tobacco may be adjusted to the changing water table, so that higher areas are planted first, at the beginning of the dry season, and lower areas later as the water table drops. A similar sequence can be seen where there is a series of river terraces at different heights, as on the Galma River near Zaria.

Regionally, there are several areas with a marked concentration of particular crops. Near to the urban centers of Kano, Zaria, and Kaduna are farmers specializing in intensive cultivation of vegetable crops for sale in the urban markets. Crops such as lettuce, cabbage, and peas are highly perishable and sold mainly to the urban population, and are rarely grown for village consumption. More remote farmers grow onions, tomatoes, peppers, and okra for local consumption and small amounts of sugar cane for chewing in the villages. They also grow crops that are less perishable, such as potatoes and rice, for eventual sale in urban markets. For example, farmers in the area southeast of Kano specialize in growing onions and peppers, and further south there is an extensive belt of *fadamas* where only sugar cane is grown. It is crushed and made into brown sugar. Most of the sugar cane is grown in broad, headwater *fadamas* just south of the Niger–Chad divide. It cannot be grown in seasonal *fadamas* and so there is very little north of this divide, with the exception of the Ikara and Palgore–Rogo areas, which have perennial *fadamas*. Around Hunkuyi, north of Zaria, farmers specialize in potatoes, and just south of Zaria, around Zangon Aya, yams are the main *fadama* and upland crop. These are grown mainly to supply the urban population of Zaria and Kaduna.

The localized production of tomatoes and tobacco near the study area is influenced by the siting of processing factories in Zaria. In 1971 the building of a tomato purée factory was accompanied by an intensive scheme for growing tomatoes in selected villages using pump irrigation. This led to a new area of specialization along the Kubani and Shika Rivers, which have a perennial water supply. Tobacco is grown in the area east of Zaria. This crop is grown both as an upland and a *fadama* crop, but recent expansion has been mainly in the *fadamas*, because Nigerians seem to prefer the flavor of the *fadama*-grown tobacco. Since the sites that are best for tobacco are also good for sugar cane, expansion of tobacco growing was limited to the areas where the *fadamas* were largely uncultivated, south of the 'sugar belt'.

Economic aspects of fadama *cultivation* In the study area, as in other parts of northern Nigeria, all farmers own upland fields on which their staple crops are grown, but the proportion of farmers owning *fadama* fields varies according to the availability of *fadama* land and accessibility to markets. In most areas, apart from those where flooding or heavy clay soils make the *fadama* cultivation particularly difficult, the *fadama* fields are of greater value than the upland fields. It is only relatively recently that land has been bought and sold for cash, but rents or sale prices for *fadama* fields are now commonly three or four times greater than for upland fields (Mortimore 1967, Norman 1972).

Yields obtained from the *fadamas* are almost always much greater than those obtained from upland fields. Figures derived from Norman's detailed economic survey of 350 farms in three villages near Zaria[3] show that net return per unit area was over three times greater and return per man–hour twice as much from *fadama* fields as from upland fields. Although *fadama* land comprised only 9.4% of the average farm area it provided 22.9% of the net farm income. Thus its economic value is far greater than its relatively small area would suggest.

The marked seasonality of the climate in the savanna regions leads to seasonal fluctuations in farm labor requirements. Harvesting of upland crops takes place during the first part of the dry season and there is a period of four or five months after this when the only work on the upland farms is repairing of huts, granaries, and fences. Dry season cultivation of the *fadamas* therefore makes efficient use of this labor surplus although there is resistance to devoting more time to *fadama* farming from some farmers who regard the dry season as a traditional time of rest, of visiting relatives, and of pursuing crafts such as weaving, basket making, and leather working.

Figure 7.2 illustrates the seasonal variations in farm labor requirements. Lines A and B on the graph show the actual number of man–hours worked on a farm by an average farming family (with labor equivalent to three adult males), plus hired labor.[4] Labor is a limiting factor in the June–July weeding bottlenock, and the labor supply at that time of year determines the amount of land that a family can cultivate. There is little hired labor available then because all farmers are busy on their own farms. Farmers who do not own *fadama* land and young men often work as hired labor on the *fadama* farms in the dry season, so labor is not likely to be a limiting factor in the dry season. Line C on the graph attempts to show the labor requirements if *fadama* cultivation is increased, assuming some *fadama* land with a perennial water supply. It shows how *fadama* farming complements upland farming, and how, if an early dry season crop dependent on groundwater is followed by an irrigated late dry season crop, the seasonal fluctuations in labor requirements can be largely evened out. The greatest labor demand on *fadama* farms is for planting (possibly April–May, September, and January,

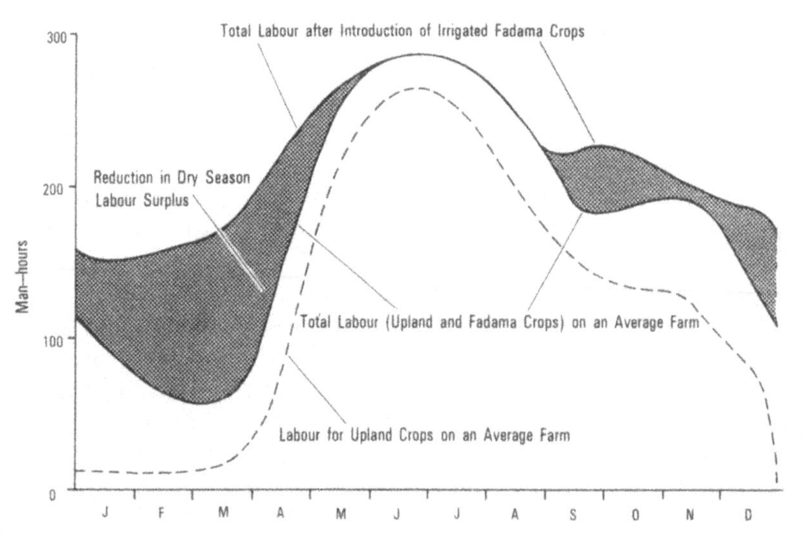

Figure 7.2 Impact of irrigation of *fadama* crops on seasonal labor requirements.

but varying according to the crops grown). Irrigating (mainly January–April) and sugar cane harvesting, which may take place at any time between late September and early March, are also demanding.

The main periods of risk of crop failure are also different for *fadama* and upland farms. With *fadama* farming, the main risks involved are sudden fluctuations of the water table and the dangers of flooding, both of which are confined to the wet season, and the danger of inadequate water supply at the end of the dry season. With upland farming, the greatest risks occur at the beginning and end of the wet season, when inadequate rainfall can cause disaster. But risks are generally fewer in *fadama* farming than in upland farming, which adds to the value of *fadama* fields in the Savanna Zone.

Fadama *grazing*

Another important aspect of *fadama* use in the study area is the relationship between the Hausa farmers and the Fulani cattle owners, often described in the past as symbiotic (see Ch. 3). The majority (over 90%) of the cattle are owned by the semi-nomadic Fulani, who migrate southwards every dry season to the area southwest of Kaduna, grazing their cattle on the *fadama* grasses, and often travelling several hundred kilometers each year, mainly because of the enormous difference between the carrying capacity for cattle in the wet season (25–77 per km^2 (55–190 per sq. mile)) and the dry season (8–40 per km^2 (20–100 per sq. mile)) in the Sudan and Northern Guinea zones (De Leeuw 1965). At the beginning of the wet season, when the upland grasses begin to sprout, the Fulani move north again, to the area north of Kano, beyond the zone infested by tsetse flies.

In the southern part of the study area, large areas are unsuitable for grazing even in the wet season because of the presence of tsetse. In the Sudan Savanna Zone, however, tsetse can only survive in the riverine areas, and eradication can be achieved simply by the clearance of vegetation along streams, whereas eradication or even control in the areas further south, where tsetse flies are very widespread, is far more complex. Extensive areas in the Sudan zone are now tsetse free and so are used for wet season grazing by the Fulani. The amount of *fadama* in these regions is, however, small, making southward migration in the dry season essential for most cattle. Eradication of tsetse in parts of the Northern Guinea zone has been attempted by barrier clearing and recurrent spraying, with some success. Reinfestation of cleared areas is a continual problem and in some cases the cost of tsetse fly eradication and control is greater than the money value of the cattle protected. Reinfestation continues to limit the amount of *fadama* available for grazing.

In the densely settled areas around Kano and Zaria, grazing is limited by other factors. Although these areas have little tree and shrub vegetation remaining in which the tsetse can live, almost all the upland is intensively cultivated annually. This means that cattle can graze during the early dry season on crop residues but the only grassland available is in the uncultivated *fadamas*, and in the late dry season and in the wet season little grazing is available. Traditionally, farmers encouraged the Fulani to graze their cattle on the crop residues because the manure fertilized their fields. The Fulani grazing areas were therefore concentrated near to settlements, where crop residues were available and where they could sell their milk and butter and buy other foodstuffs. However, the expansion of *fadama* cultivation increased, thereby reducing the amount of *fadama* available for dry season grazing, and this gradually resulted in overgrazing of *fadamas* and other pastures near many towns and larger villages, resulting in destruction of vegetation and soil erosion. Damage to *fadama* crops by cattle trampling and eating them, combined with the competition for land in these densely settled areas, have resulted in deteriorating relationships between farmers and cattle owners, resulting sometimes in court action.

Attempts have been made by the government in recent years to persuade the Fulani to settle in areas that have been designated as grazing reserves. The location of these areas is, however, far from ideal from the Fulani point of view and so far little success has been achieved. Many of the grazing reserves have no perennial supplies of water and contain very little *fadama* to provide dry season grazing. Moreover, they are remote from Hausa settlements and especially markets, so the Fulani would be unable to sell their milk or buy other foodstuffs as they have done traditionally. Better siting of future reserves may meet with greater success, and the abolition of the cattle tax[5] in 1975 removed one of the basic deterrents to Fulani settlement, but it is unlikely that all of the nomadic Fulani will want to settle permanently.

Land use and technological development

The land use of the *fadamas* may be classified according to the increasing complexity of technology employed. Each level of technology and associated land use is discussed in turn.

Use of unmodified fadamas At the simplest level, the unmodified vegetation of the *fadamas* is used to provide edible and medicinal plants, and material for weaving mats, baskets, and fences. The sand and clay found in *fadamas* are used for building. The small animals that live there are hunted and the water is used for domestic supplies and for watering livestock. This level of land use also includes the use of *fadamas* for grazing and fishing. These uses are all widespread among peoples of northern Nigeria and are restricted only by the availability and size of the *fadamas* and, in the case of grazing, by the distribution of tsetse flies and the extent of flooding.

Traditional agriculture The second level of land use involves cultivation of the *fadamas* using hoes and groundwater resources. In the dry season this is possible only where the groundwater level is within reach of crop roots, which means that some *fadamas* can be used during the first part of the dry season but become too dry during the late dry season. On the other hand, others are too wet at the beginning of the dry season and can only be used when the surface water has dried out. In the wet season, cultivation is often severely restricted by waterlogging and flooding, but farmers alleviate these problems by planting their crops on mounds or ridges, which may be up to 1 m high, depending on local conditions.

The amount of capital needed to invest in technology for this type of agriculture is very small and the majority of cultivated *fadamas* in northern Nigeria are farmed at this level of technology. Almost all *fadamas* are suitable for this type of farming for at least part of the year, although the choice of crops may be limited.

Improved agriculture The third level of land use involves imported technology and improved agriculture. Improved agriculture on *fadamas* is achieved in three basic ways:

(a) use of 'improved techniques' not involving mechanization;
(b) water control (irrigation and, less commonly, drainage);
(c) mechanization for cultivation (plowing, sowing, etc.).

These developments are not mutually exclusive in time or space, but they are considered separately here for the sake of clarity.

USE OF 'IMPROVED TECHNIQUES' NOT INVOLVING MECHANIZATION Improved techniques are of two kinds: the first includes changes in the dates of planting, in the spacing of plants, and the use of rotation and alternate cross tied furrows or mulches to conserve water. These techniques do not require the purchase of any costly inputs and so are within the reach of any farmer. The second group of techniques includes the use of fertilizers, herbicides, pesticides, and improved seed varieties, which do require some investment capital to purchase inputs but which, nevertheless, should be within reach of most farmers with the help of government loans, although there are often other difficulties such as obtaining required supplies, particularly in remote areas.

Most importantly, the use of these techniques is not usually restricted by site factors, although soil factors affect fertilizer application and it may not be easy to obtain advice on requirements for the variable *fadama* soils. The main reason why these techniques are not more widely applied at present is the lack of knowledge among farmers. This problem could be remedied by improvements in the agricultural extension service.

WATER CONTROL Shadoofs are the most usual method of irrigation. Small areas are irrigated by drawing water from streams or shallow wells in buckets or calabashes. The water is poured into small channels and the fields are irrigated by gravitational flow. The areas suitable for shadoof irrigation are restricted by availability of water near the surface. Sometimes canals are dug from a stream channel to a shadoof site but these are rarely more than 20 m (66 ft) long and are usually built from a channel that occupies only a small part of wide river bed during the dry season to the river bank. The maximum height of lift of a shadoof is about 3–4 m (10–13 ft).

The materials used for shadoof construction are universally available: clay, poles, calabashes, and rope. Construction is widely known and costs are very low. The amount of labor needed for shadoof irrigation is, however, considerable. The area to be irrigated must be laid out so that there is a slight slope away from the shadoof, and a network of channels to lead the water to all parts of the plot. The irrigation itself involves several hours hard labor at intervals varying between 2 days and a week. Usually two men are involved, one to operate the shadoof and the other to control the flow of water to the various channels, using a hoe. A recent development is the use of sheets of polythene to line the first part of the channel to prevent water loss. The area which one man can cultivate using a shadoof is less than that using groundwater because of the extra labor involved, but yields are generally much higher and the length of the growing season is often extended by use of irrigation.

The use of diesel pumps for irrigation is becoming more common. A much larger area can be irrigated with a pump than is the case with a shadoof: approximately 4 ha (10 acres) compared with 0.1 ha (0.25 acre). The water can also be lifted to a greater height. In addition, water can be pumped to an area not immediately adjacent to the water supply. This is the case at the Galma Scheme near Zaria, where the terrace is irrigated but not the floodplain. This means that sites suitable for cultivation, but which could not be irrigated by shadoof, can now be brought into production. On the other hand, small sites, or those with a limited supply of water, remain more suitable for shadoof irrigation and pumps should not be employed. Possibly the main advantage of pump irrigation is its low labor feature, although it is still necessary to direct the water into the correct channels by hand.

On the other hand, there are important disadvantages in pump irrigation. First, the initial cost of the pump is too high for many farmers to purchase individually, requiring that they either cooperate in purchasing a pump (with the help of loans) or that government or other organizations arrange a hiring scheme. Secondly, there is the continuing expense of fuel and maintenance. There is often a shortage or a complete lack of diesel fuel in rural areas and farmers must travel long distances to obtain it. Thirdly, in the case of mechanical breakdowns, there are great difficulties in obtaining the services of a diesel mechanic, and in obtaining spare parts. It may take several months to repair a pump. This is a major problem because if crops are left without water for more than a few days, they will die. It is not possible to water by hand because of the large area involved.

Crops that are grown for factory processing are needed in relatively large quantities and so are often particularly well suited to pump irrigation, as for example wheat and tomatoes, both of which can be grown only in the dry season. The profits obtained from pump irrigation schemes are much higher than from smaller scale shadoof irrigation. One group of farmers who had combined their resources to rent a pump from the Ministry of Agriculture to grow dry season vegetables reported profits five times greater than with their previous shadoof irrigation.[6] However, as pointed out, although profits are much greater, the accompanying risks are also much higher.

The construction of small earth dams may be considered in this level of technology. There are not many in the area at present, except on government experimental farms (e.g. Samaru, Maigana), but more are being constructed either in conjunction with pump irrigation schemes or, in some cases, mainly to provide water for cattle and other livestock.

Large-scale irrigation schemes, involving construction of concrete dams and canals, are not confined to *fadama* sites and may involve irrigation of considerable areas of upland as in the Kano Irrigation Scheme. The other large-scale scheme in the area (the Kangimi Scheme) involves irrigation of the terraces of the Kaduna River.

Schemes such as these require plentiful supplies of water and are therefore confined to major rivers. The sites that are suitable are further restricted by the necessity for suitable dam sites. The capital investment needed is very large and such schemes are almost always financed through the government or through international agencies.

After the initial construction of the irrigation structures, the level of mechanization involved varies considerably. The Kano Scheme involves tractor-hiring for land preparation, but most of the other farm operations are carried out by individual farmers using hoes. Other schemes may involve a much greater degree of mechanization, using combine harvesters, etc., but this is found only in the case of schemes producing crops for commercial processing. Generally, availability of labor is not a limiting factor and so mechanization is likely to develop slowly.

MECHANIZATION FOR CULTIVATION Mechanization involves the use of tractor-drawn equipment which is useful for leveling, clearing, and plowing land before planting, but it is not usually feasible for operations such as weeding and harvesting without completely changing the methods of farming and attitudes of the farmers. Consequently, these activities are done manually. In addition, the small size of many *fadamas* and the small, fenced or hedged fields preclude the use of tractors in many cases. Machines are most easily used on larger *fadamas*, especially those being cultivated for the first time. The restrictions due to cost of fuel and spare parts apply as in the case of pump irrigation and cooperative purchasing or hiring schemes are necessary for most farmers.

Larger, more complex, machinery such as combine harvesters is only found on a few government irrigation schemes. It can only be used on the larger floodplain *fadamas* where large areas are devoted to a single crop in an organized scheme.

The potential for development

The variations in physical conditions between different *fadamas* allow for many different uses, and the productivity of farming, grazing, and fishing can all be considerably improved. In many cases, multiple use of the *fadamas* is possible, either simultaneously, as in the case of fishing and cultivation, or at different times of the year. Some *fadamas* are not suitable for cultivation and others may be used for groundwater cultivation but do not have sufficient water for irrigation. Usually, the higher the stage of technology, the more commercially oriented the farming is; greater profits can be made, but fewer sites are physically suitable and greater problems and risks are involved.

Most *fadama* land can be cropped for at least half the year with very little modification, and much of it can, with minor modifications, be cropped or grazed throughout the year. Development of the *fadamas* can take place in three ways:

(a) extension of the area used;
(b) intensification of present use;
(c) changes towards higher yields or higher value crops.

Extension of area used In the study area, *fadamas* covered 185 000 ha (457 000 acres) and of this, 88 000 ha (218 000 acres) were uncultivated in 1962.[7] This gives one indication of the potential for development, but it is neither possible nor desirable that all *fadamas* should be cultivated. The nonagricultural uses of the *fadamas* are also important, especially fishing and grazing. The construction of dams results in loss of grazing land, which is particularly serious when it is near the towns, as in the case of the Galma Dam near Zaria (Fig. 7.3). Constructed in 1975, the Galma Dam resulted in the flooding of 1659 ha (4100 acres), most of which was formerly used for grazing. It is likely that in the future more fodder crops will be grown in the *fadamas*, but this will not eliminate the need for substantial areas of grazing.

Presently, most of the unused *fadamas* are too wet or too remote. Extension of the area cultivated is bound to take place as the population expands into underpopulated areas and will be accelerated by the

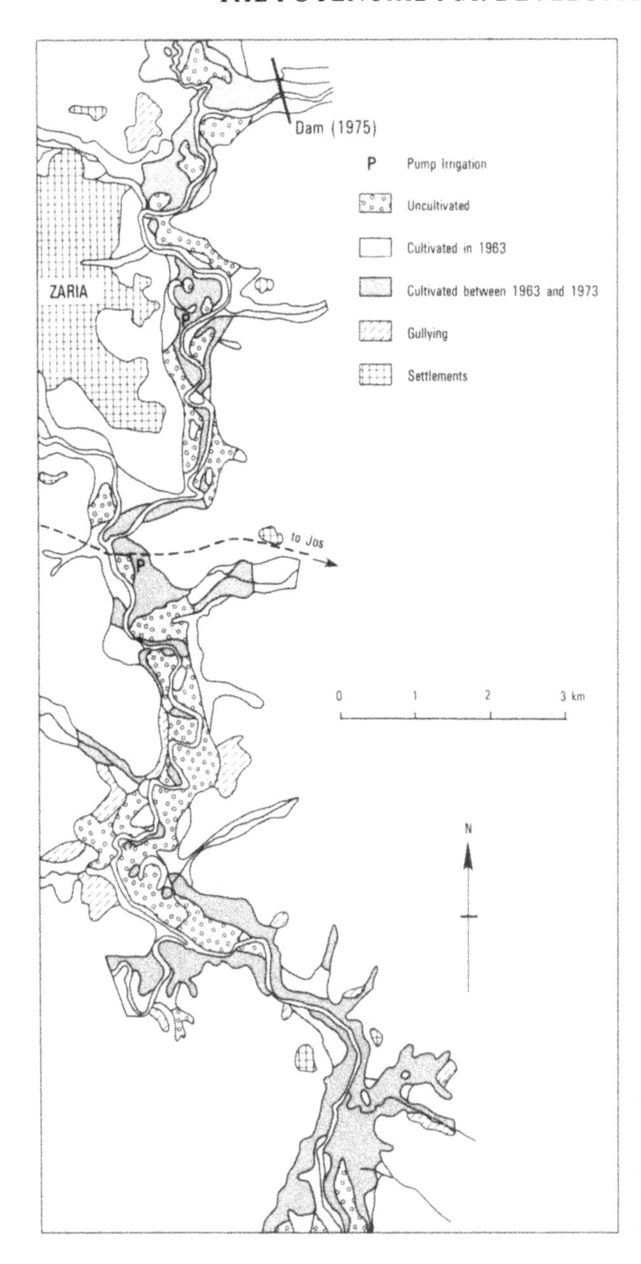

Figure 7.3
Expansion of *fadama* cultivation on the Galma flood plain near Zaria, 1963–73.

development of irrigation projects, such as the proposals for construction of a series of small dams on the tributaries of the River Galma east of Zaria (Fig. 7.3).[8] At the higher levels of technology, flood control and wet season drainage of the floodplain *fadamas* will allow expansion of both grazing and cultivation.

Intensification of use In most *fadamas*, multiple usage is possible. In the *fadamas* with perennial streams, increased fish production would result from fish breeding and stocking, which is little practiced at present. The building of small earth dams close to villages will provide opportunities for year-round fishing, water supply for household use and domestic animals, and although cultivable areas may be inundated, small-scale irrigation of new areas should become possible to compensate for the areas lost. Such dams would allow some cattle to remain throughout the year, supplying manure for the crops, and milk and milk products for the villagers.

Crop yields can be increased in two ways, either by using the *fadamas* for a greater part of the year (i.e. double or treble cropping), or by use of the 'improved techniques' mentioned above. Many *fadamas*, which are at present used only in the early dry season, could be used in the late dry season, either by irrigation or by planting deeper-rooting crops, and in the wet season by growing rice.

Where shortage of labor is a factor in limiting the introduction of irrigation, the use of animals for drawing water may be possible. This is a common practice in Niger and Sudan, but it is not employed in Nigeria. Mixed farmers often use the animals for plowing upland fields at the beginning of the rainy season, and occasionally for drawing carts, but they could also be used on the larger *fadamas* both for drawing water and pulling other implements such as harrows and seed drills. Growth of fodder crops, making of hay and silage, and feeding animals on the residues from crop processing (such as groundnut cake) would greatly increase animal productivity.

Although many of these improvements can be undertaken by individual farmers, others require group cooperation or government backing for loans, tractors, pumps, and for the purchase of fertilizers, pesticides, and seeds. Other government involvement is needed in conducting more research into improved varieties of *fadama* crops, ensuring adequate supplies of fertilizers, diesel fuel, etc. and in expansion of extension services to provide better information and demonstrations of techniques to the farmers.

CHANGE IN CROPS In some cases it may be possible to improve yields and to use the *fadamas* for a greater part of the year by changing to crops that are better suited to the physical conditions in particular *fadamas*. Variations in water table levels are of critical importance, and adjustment to crops with rooting depths best suited to the water table conditions can be beneficial. Most crops also have quite a narrow range of optimum climatic conditions, although tolerating a much wider range of other conditions (Chang 1968, Black 1971). Much more research is needed into these factors, particularly for *fadama* crops under the prevailing climatic conditions.

Limitations to development

The physical conditions that limit use of the *fadamas* may be summarized as regional conditions, including amount of rainfall, length of the rainy season, prevailing temperatures, and incidence of pests and diseases, and local conditions, including size of *fadama*, amount of surface water, depth of groundwater at different times of the year, and soil conditions. As explained above, adjustments can be made in the use of *fadamas* and in *fadama* crops to derive the greatest benefit from the prevailing physical conditions. There is great variability in local conditions from one *fadama* to the next and therefore in the optimum use. There is also the variation from year to year in the hydrological conditions within the *fadamas*, and so crop adjustments may have to be made from one year to the next.

Other problems arise from misuse of both the *fadamas* and the adjacent upland areas. Clearance of natural vegetation, either for expansion of cultivation, to supply firewood, or by burning for hunting or grazing, has a drastic effect in increasing soil erosion, in causing changes in the amount of runoff, and in the groundwater levels (Carter & Barber 1958, Jones 1960). In the Sokoto and Rima valleys, northwest of the area studied, clearance of vegetation in the headwater regions has resulted in increased flooding, increased sediment load in the rivers, and decreased flow in the late dry season, and has led to the abandonment of many *fadama* farms (Ledger 1961). Gully erosion has also been caused in the study area by the destruction of vegetation, by overgrazing, and by the overuse of cattle tracks and footpaths, particularly where they cross streams. Some farming practices, such as construction of ridges at right angles to the contours and exposure of bare earth to wind and water erosion, also lead to gullying and sheet erosion. The seriousness of the problem can be seen from the fact that 6% of the study area within the Chad basin has been destroyed by gully erosion.

Technological factors also limit development and show considerable spatial and temporal variation. The importance of site factors such as the increasing amount of water and increasing size of *fadama* needed at the higher levels of technology has been mentioned. Some factors such as waterlogging or danger of flooding may be more of a problem to an individual farmer involved in groundwater cultivation than at a more advanced level of technology where preventative measures can be taken. But usually at the more advanced levels of technology the limitations increase. Irregular topography matters less to the farmer relying on groundwater than it does in irrigation schemes where leveling is necessary to ensure even distribution of the water. Similarly, variability of soils can be coped with by a farmer cultivating small plots of different crops, but it poses greater problems on a large-scale scheme. If high-yielding varieties of crops are introduced, it is essential that correct amounts of fertilizer and pesticides are applied, otherwise yields will be lower than with hardier traditional varieties.

Socio-economic factors also provide constraints on development. The problems of obtaining supplies of fertilizers, pesticides, and diesel fuel, and the difficulties in repairing mechanical breakdowns have been mentioned. The attitudes of the farmers, particularly with regard to motivation, are also important. Most farmers are not primarily motivated by considerations of profitability, but regard security as of greater importance. They ensure that their food crops are planted and cared for before attending to their cash crops. Similarly, although desire for material benefits may motivate them to plant cash crops, the suspicion that extra cash will only be taken from them in extra taxes and the desire for a time of leisure in the dry season, may deter them from producing much beyond the immediate needs of their family (Norman 1972). In the larger schemes, dislike of enforced regulations and loss of personal freedom are also deterrents.[9]

Labor may become a limiting factor at the more advanced levels, because to justify the capital expenditure, it may be necessary to cultivate the *fadamas* in the wet season as well as the dry season, and so competition with labor requirements for upland crops will become a problem (see Fig. 7.2 and p. 156). Economic problems result from the increasing cost of *fadama* land, which is likely to accelerate since rapid population growth is leading to land shortage in many areas. The difficulties of access to markets also restrict development in the *fadamas* remote from road or railway. This is true because the most profitable *fadama* crops are either vegetables, which are highly perishable and cannot be grown unless rapid transport to market is assured, and sugar cane, which must be crushed within 24 h of cutting, which in turn means it requires rapid transport to the nearest mill. Sugar cane is also a very bulky crop and transport costs are high.

The limitations to development are therefore extremely varied but many of them apply only to certain areas or to certain types of development. At any particular *fadama* site any of the levels of technological development may be optimum and so all levels are likely to continue to exist side by side.

Regional differences in development potential The potential for development also varies considerably from one region to another. To analyze this, the study area is divided into two contrasting regions separated by the Niger–Chad divide shown in Figure 7.1. To the north of this divide there are fewer *fadamas*, they are more seasonal than perennial, and a much greater proportion are already cultivated (Table 7.1). There is also a higher population density and a much greater area suffering from gully erosion. In this region expansion of cultivation is limited by the small amount of uncultivated *fadama*. Nevertheless, extension of irrigation and use of the *fadamas* for a greater part of the year is possible in some cases, but is limited in many *fadamas* by lack of water. The extensive gully erosion leads to rapid

Table 7.1 Contrasts across the Niger–Chad divide.

	Niger Basin[a]	Chad Basin[a]	Level of significance of difference (%)[b]
gully erosion (mean % of area)	1.78	6.02	0.1
drainage density (km per km^2)	0.78	1.10	5
stream frequency (no. channels per km^2)	0.92	1.86	0.1
total *fadama* (mean % of area)	14.78	4.51	0.1
seasonal *fadama* (mean % of area)	0.94	2.35	0.1
perennial *fadama* (mean % of area)	13.83	2.16	0.1
width of *fadama* (km)	0.44	0.17	0.1
proportion cultivated (mean % of total *fadama*)	29.35	86.96	0.1
n for proportion cultivated	211	129	
n for all other variables	289	245	

The variables were measured in randomly selected quadrants, each 1 km^2.

[a]These terms refer to the parts of the drainage basins falling within the study area (see Fig. 7.1).

[b]Two-tailed tests in both cases.

runoff and earlier drying up of the streams. Most gullying is still active and conservation measures are urgently needed to prevent further deterioration. A recent problem affecting the potential development of this area is the large amount of water now being removed from the Kano River at the Tiga Dam for the Kano Irrigation Scheme. This reduces the amount of water available down stream and thus severely limits the agricultural possibilities. Most *fadama* development in this region will therefore be restricted to increasing the yields of individual small-scale farmers.

South of the Niger–Chad divide conditions are very different. There is much more *fadama*, most of it is perennial, less is cultivated, and there is much less gully erosion. Most of the cultivation is concentrated around Zaria, and along the Zaria to Kaduna and Zaria to Jos roads. There is considerable potential for expansion of the area of *fadama* cultivated and for further development of irrigation. The greater availability of water and the larger size of the *fadamas* allow cultivation of most *fadamas* throughout the year, and use of mechanization in many areas. There is also scope for improvement of fishing in many of the streams. The towns of Zaria and Kaduna and many large villages have well established markets and it is likely that development will take place initially close to these markets or to the existing roads and railway, and will expand later into the more remote and less densely populated areas.

Factors limiting development include the presence of tsetse which restricts grazing in part of the area, and extensive forest reserves in which cultivation is forbidden. The low settlement density in the east of the area and the rather sparse network of roads and railways may slow down or limit expansion in these areas and the variability of physical conditions in the large floodplains of the Galma, Kaduna, and Karimi Rivers restricts development.

In conclusion, there is potential for further development in both parts of the study area, but it is much greater to the south than to the north of the Niger–Chad divide. Careful planning of future development is necessary to ensure the optimum benefit for farmers and to minimize friction between farmers and herders who are now competing for use of the *fadamas*. In addition, detailed studies of morphology, hydrology, and soils are needed to provide a sound base for *fadama* land-use recommendations. Further research is also needed into the effects that changes in land use have on the hydrology and soils and into the possible physical and socio-economic impact of any proposed developments.

The study area is typical of the Savanna Zones throughout Africa in all but one respect: that the northern area around Kano is much more densely populated and intensively used than almost any other Savanna area. The conclusions for the southern part of the area are therefore more widely applicable and there is clearly considerable potential for further development of the *fadamas* in many parts of Africa.

Fadama cultivation can increase food supplies, improve nutritional standards by adding variety to the diet, and reduce the disastrous impact of the failure of rainfed crops. By increasing the opportunities for dry season labor and for cash-crop production, *fadama* cultivation proves an ideal method of developing rural areas and reducing rural–urban migration

Notes

1 *Fadama* is a Hausa word which has become widely used in the geographical literature on West Africa. It is roughly equivalent to *dambo* or *vlei* in East and South Africa. The correct plural form in Hausa is *fadamu* or *fadamoni* but '*fadama* lands' or '*fadamas*' are the terms usually used in the literature.

2 *Editor's note:* today, demand for rice is outstripping domestic supply and in 1980 Nigeria had to import many hundreds of pounds of rice.

3 Figures derived from the data given in tables 8, 65, and 73 (Norman 1972) show that average farm size was 3.95 ha (9.75 acres), of which 3.57 ha (8.83 acres) were upland and 0.37 ha (0.92 acres), or 9.44%, *fadama*. The figures also show that 0.67 ha (1.64 acres) or 18.62% of the upland and 0.08 ha (0.20 acres) or 22.04% of the *fadama* was fallow. These proportions varied considerably between the three villages and it was only in the most inaccessible village that a greater proportion of *fadama* than upland was left fallow.

The average net return per hectare was 549.9 shillings for upland (222.5 shillings per acre) and 1607.7 shillings (650.63 shillings per acre) for *fadama*, excluding labor costs. Compare 499.4 and 1485.8 shillings per hectare (202.1 and 601.3 shillings per acre) costing only nonfamily labor, and 275.3 and 945.3 shillings per hectare (111.4 and 382.6 shillings per acre) costing all labor.

The net annual farm income from upland was therefore 1448 shillings and from *fadama* 431 shillings (costing nonfamily labor).

4 Based on data given in a study of 350 farming families in three villages near Zaria that have varying amounts of *fadama*.

5 Cattle tax (*jangali*) was levied according to the number of cattle owned but counting was often evaded by the migrating herdsmen.

6 Farmers reported profits of over N40 per hectare (N100 per acre) in 1973 from 4 ha of vegetables.

7 These figures are derived from air photographs, which were the most recent photographs available at the time of the study. Others taken in 1973 are now available.

8 Plan by the Kaduna State Government in 1976.

9 Norman mentions other reasons that deter farmers from additional work over and above subsistence needs (Norman 1972). These include the custom of sharing extra income among the extended family, physical health, attitudes to farm work, educational level, and climate.

References

Africanus, L. 1600. *The history and description of Africa* (trans. J. Pory 1896), 828–31. London: Hakluyt Society.

Andrews, D. J. 1968. Wheat cultivation and research in Nigeria. *Nigerian Agric. J.* **5**, 67–72.

Black, C. C. 1971. Ecological implications of dividing plants into groups with distinctive photosynthetic production capacities. *Adv. Ecol. Res.* **7**, 82–114.

Carter, J. D. and W. Barber 1958. The rise in the water table in parts of Potiskum Division, Bornu Province. *Recs Geol. Surv. Nigeria* **3**, 5–13.

Chang, J.-H. 1968. *Climate and agriculture: an ecological survey*, 75. Chicago: Aldine.

Dalziel, J. M. 1937. *The useful plants of West Tropical Africa*. London: Crown Agents.

De Leeuw, P. N. 1965. The role of savanna in nomadic pastoralism: some observations from Western Bornu. *Netherl. J. Agric. Sci.* **13**, 178–89.

Extension and Research Liaison Service 1971. *Rice production in northern states of Nigeria*. Exten. Bull., no. 8. Institute for Agricultural Research, Ahmadu Bello University, Zaria, Nigeria.

Hogben, S. J. and A. H. M. Kirk-Green 1966. *The emirates of Northern Nigeria: a preliminary survey of their historical traditions*, 222. London: Oxford University Press.

Johnston, H. A. S. 1967. *The Fulani Empire of Sokoto*, 157. London: Oxford University Press.

Jones, D. G. 1960. The rise in the water table in parts of Daura and Katsina Emirates. *Recs Geol. Surv. Nigeria* **4**, 24–8.

Kirk-Green, A. H. M. 1962. *Barth's travels in Nigeria*, 113. London: Oxford University Press.

Last, M. 1967. *The Sokoto Caliphate*, 184. London: Longman.

Ledger, D. C. 1961. Recent hydrological changes in the Rima Basin, northern Nigeria. *Geogr. J.* **127**, 477–87.

Mortimore, M. J. 1967. Land and population pressure in the Kano close-settled zone, northern Nigeria. *Advancement Sci.* **23**, 677–86.

Norman, D. W. 1972. An economic survey of three villages in Zaria Province. 1. Land and labour relations. 2. Input–output study, vol. i, Text, vol. ii, Basic data and survey forms. *Samaru Miscell. Paps 1* **19**, 37 and 38.

Turner, B. 1977. *The* fadama *lands of central northern Nigeria: their classification, spatial variation, present and potential use*. PhD thesis. University of London.

8 *The Fulani in a development context: the relevance of cultural traditions for coping with change and crisis**

PAUL RIESMAN

The Fulani of West Africa are one of the major cattle-raising peoples of the world. They are unique among African peoples in the degree to which they are scattered over a vast area. Not only are there significant groups of them in every state of West Africa, but also there are Fulani living in Cameroon, Chad, the Central African Empire, and the Sudan. Over half the Fulani do not raise cattle today, but there are millions who do and they are concentrated for the most part in the poorest countries of West Africa. This is significant, because the exporting of meat can become a major source of external revenue for the state. Governments are under pressure to increase meat production because of the dual need to make more meat available for the domestic market and to improve their balance of payments situations through export to overseas markets. Thus in all the countries where the Fulani are significant minorities the governments are hoping to increase meat production and sale by encouraging economic development projects among the Fulani.

Economic development projects among livestock raisers must be examined from both a moral and a practical standpoint. On the one hand, the issue is whether the procedures used to try to bring about development are moral. In other words, do they end up treating the people with respect for their integrity? On the other hand, do the projects really work in practice?

A major snag in many overseas development projects has been the human factor. All too often development experts fail to take social structure, way of life, and world view of the people they hope to help into account. This inevitably results in one or more of the following difficulties: (a) radical misinterpretation by the 'target' population of what the experts are trying to do; (b) significant disruption of a continuing way of life, with the benefits

*This work was developed from a report produced for the USAID Sahelian Social Development Series under contract number REDSO/WA78–138.

going just to a few rather than to the majority or to all; (c) deterioration rather than improvement of the economic situation due to the manner in which the target population makes use of the facilities or other changes created by the project. This list could be expanded.

The major aim of this chapter is to help reduce these kinds of difficulties in the case of the Fulani and whatever development projects might be tried among them. Once we have made a commitment to acting morally – e.g. renouncing the use of force – then our hope of reducing these difficulties rests on achieving a deep understanding of the Fulani way of life as a total system, and on correctly analyzing how changes introduced at various points in the system might affect its overall functioning. My goal here is to describe this system and to argue that the main features of the Fulani way of life are not whimsical and mysterious products of an exotic mentality, but that they are the logical results of the interplay between known human needs, known animal needs, and constraints in the geographical and social environments.

The Fulani

The Fulani are one of West Africa's numerically largest groups. With a total estimated population of 9 or 10 million, they number in the same order of magnitude as the Hausa, Igbo, and Yoruba of Nigeria. They do not form a majority in any West African state, but they are very large minorities in all the Sahelian states, as well as in Guinea, and Fulani groups dominate particular regions within those states. Table 8.1 gives the figures for the Fulani population of most of the states where they live.

There is a major division between two life-styles lived by the Fulani: the pastoral and the sedentary. The former includes both nomadic and semi-sedentary modes. Nomadic and semi-sedentary pastoralists frequently shift back and forth from one to the other variant of the cattle-herding life, but once Fulani have given up cattle raising for some other occupation they rarely go back to it again.

The origin of the Fulani is the subject of much speculation, but modern linguistic and historical research connects the Fulani with the banks of the Senegal River some time during the first millennium of our era. (For a detailed discussion of their origin and settlement, see Ch. 3.)

It is important to point out, however, that linguistic evidence supports the hypothesis that the Fulani expansion was both fairly recent and fairly rapid. The Fulani language, called Fula by linguists and Fulfulde by the Fulani, is mutually comprehensible by speakers from nearly all areas. There are exceptions, especially where small enclaves of Fulani are surrounded by speakers of another language. A good example is in Barani, northwestern Upper Volta, where Fula is heavily influenced by the Marka language.

Table 8.1 Fulani populations[a]
(a) *Breakdown by country*

Country	Year of census/source	Number
Benin	1952	54 000
Cameroon	1960–1	400 000
Chad	1964	32 000
Gambia	1954	58 700
Ghana	1950	5 500
Guinea	1970	1 500 000
Guinea-Bissau	1948	36 500
Ivory Coast	1952	52 000
Mali (estimate based on cattle population)	1975	400 000
Mauritania (Peul)	1962	40 000
Mauritania (Toucouleur)	1962	70 000
Niger	1962	247 143
Nigeria	1972	4 800 000
Senegal (Peul)	1969	560 000
Senegal (Toucouleur)	1969	442 000
Upper Volta	1974	300 000
TOTAL		8 997 843

(b) *Breakdown by mode of life* (see pp. 174–8)

	Pastoral	Sedentary	
nomads	93 300	5 886 400	
semi-sedentary	3 018 143		
TOTAL	3 111 443	5 886 400	8 997 843

[a]These data are for different dates and are reasonable estimates. This is because of the various ways in which Fulani are defined.

It is also important to make clear the distinction between certain terms used in the Fulani language because they have significant meaning in contemporary society. The most important of these is the distinction between noble and slave, *pullo* (pl. *ful'be*) and *maccu'do* (pl. *maccu'be*), because it has been basic to the shaping of Fulani society as it exists today.

This distinction also helps us to understand why the non-*Ful'be* strata of Fulani society, especially the *maccu'be* and the *riimaay'be*, may have the most to gain from political and economic development. The Fulani have good reasons to want to conserve their position in society, but the *maccu'be* have good reasons to want to change theirs. Thus it is perfectly logical that the 'conservative' Fulani generally have rejected modern education, whereas the 'progressive' *maccu'be* have sought it. Ironically, this education has now made it possible for some *maccu'be* to return home as government officials and rule their former masters. We may permit ourselves to smirk momentarily at this 'logical' response of the Fulani, but this is precisely the kind of behavior that needs to be looked at more closely. We must not jump to the conclusion that we understand it simply because it makes a neat contrast to the way the *maccu'be* have acted. If we do jump to such a conclusion, we in effect close the door to further understanding because we stop trying, and we thereby curtail our effectiveness as agents of beneficial change.

How the Fulani live

The Nomads Very few Fulani today are true nomads – that is, people who move from place to place with no fixed home and who live entirely off the produce of their herds. In contrast to some East African pastoralists, such as the Masai (Jacobs 1975), no Fulani group, to my knowledge, lives day in, day out, on the products of its herd alone. The Fulani do not make use of the blood of their animals to drink or in any other way, and though at certain times of the year they may live on milk alone, millet is the staple of their diet almost everywhere in West Africa. It is a basic given of nomadic Fulani economy, then, that there is trade with agricultural peoples.

There are two major points to remember concerning the economic base of nomadic life when looking at it in a development context: (a) the first is that a group's herds are made up such that, in normal times, the people can satisfy all their subsistence needs on the revenues from milk products alone. No nomadic Fulani are, or have ever been, oriented toward the production of beef. Cattle are of course sold from time to time, but from the point of view of the herd manager these sales are generally in response to large financial needs above and beyond subsistence, such as medical expenses, taxes, ceremonial expenses, clothing and, lastly, cooking utensils and manufactured items. (b) The second major point is that the nomads and the sedentary peoples among whom they move are, in a certain sense, a luxury for one another. The nomad economy does depend absolutely on the products of the agricultural economy, but the nomads are such a small minority of the population that the amount of millet they obtain from the farmers is negligible when compared to their total harvest. In exchange, the milk products that the farmers buy or trade from the nomads are luxuries in their own diet. Only the richest of villagers, such as shop-owners, traders, or civil servants would be likely to buy milk or butter every day. The interdependence of these peoples is thus not at all like what we are familiar with in a modern, market economy.

Nevertheless, exchange is essential (also see Ch. 3). Recent detailed surveys (Dupire 1962, Dahl & Hjort 1976) and my own research show that only a minority of the well-off households can live on their herds' milk alone and this can be done for only a few months during the height of the rainy season. During the rest of the year, food must be obtained some other way, usually through exchange with agriculturalists for millet.

Though nomads rely on exchange with agriculturalists to obtain millet, it is unlikely that they will become sedentary farmers. A significant reason for this is that cattle herding turns out to be an efficient way of exploiting an environment characterized by a short, uncertain rainy season and a long dry season. It is these same factors, however, that make life hard for herders, both nomadic and sedentary. The dry season is an unpleasantly hot time for everybody in the Sahel. But this is the very period that herdsmen must do their most demanding and grueling work, watering the cattle and taking them to and from pastures of dry grass. This is obviously hard work, but it is made even harder as the dry season progresses. For example, toward the end of the dry season a man may easily spend night and day watering his cattle. In addition, as the cattle eat up the grass in the vicinity of the well they have to be taken farther and farther away to get adequate food.

These inevitable difficulties and hardships of keeping a herd alive are among the factors that place upper limits on the number of animals a man can take care of and thereby they actually play an important role in the success of the nomadic way of adapting to the environment.

The adaptive mechanisms of the Fulani are discussed in Chapters 1 and 3, but it is important to point out here that the way land, water, and animals are owned is especially suited to the Sahel environment. For example, among all the nomadic peoples of the Sahel it is generally felt that pastureland belongs to no one and that anyone therefore has a right to use it. At the same time, everyone recognizes the good sense of not pasturing cows near wells in the rainy season and thereby needlessly using up valuable dry season pasturage. Similarly, among nomads, natural sources of water such as streams, ponds, and waterholes are not considered to belong to individuals or groups. Because grass is found nearly everywhere during the rainy season, herders usually travel in small groups, spreading out over the landscape and thereby taking maximum advantage of fresh grass as it becomes available.

Wells, however, are an entirely different matter. Wells are owned by the person or group that dug them or had them dug. The need for well-ownership in order to obtain water, and the fact that someone is going to have to work very hard in order to quench the thirst of a whole herd, are thus factors that help keep an upper limit to the size of herds and thereby induce people to spread the cattle around as much as possible in the dry season. The 'spreading around' of cattle, then, maximizes use of the land without degrading it (for more on this point, see Ch. 5).

Historically, however, optimum utilization of the land has not always been people's prime consideration. In fact, I would argue that it is only during the colonial and modern periods that such economic factors have been able to have free play. This idea is supported by the observation that most of the nomadic Fulani are newcomers to the regions they are found in today (Dupire 1962, p. 51; Horowitz 1972, p. 106). One of the reasons they have been able to infiltrate zones that were traditionally controlled by the Tuareg (cf. Dupire 1962, pp. 91–2) is probably that the once powerful Tuareg society can no longer control its own territory by force of arms, nor frighten small bands of herders with the threat of violently stealing their cattle. In general, the dangers of violence and cattle theft, which had formerly been major influences on the ecological adaptation of Sahelian peoples, have been almost negligible in most areas for two to four generations. The effect of those dangers in the past would have been greatly to limit freedom of movement, to counteract the extreme tendencies to split up into small camps, and to slow down penetration into unexploited areas of the Sahel. All of these effects together probably made the growth rate of both human and bovine populations lower than it is today. Another consequence of the violence of the precolonial period, however, was that some semi-sedentary Fulani took better care of their cattle then than they do today. As we shall see in the next section, one of the ways the semi-sedentary life-style differs from the nomadic one is that constant surveillance of the cattle is not necessary for the former because much of the year they are in 'home territory'. The nomads, on the other hand, really have no home territory and must therefore be on their guard most of the time.

The non-nomadic populations It is the non-nomadic Fulani who make up the vast majority of the total Fulani population, since of the total of 9 million or more only about 100 000 are pure nomads. For our purposes here it will be useful to classify the non-nomadic Fulani according to both social and geographic criteria into three main groups: (a) the 'semi-sedentary', which includes all Fulani who farm and for whom cattle-herding is also important, regardless of how much they actually move around; (b) the 'sedentary', which includes Fulani who farm or pursue other occupations, and who do not do a significant amount of cattle herding; (c) the religious elites, who form a class apart and who are supported by Fulani (and to some extent other) Muslims from all walks of life. Within my category of semi-sedentary Fulani there is, however, a significant geographical difference in that some of these Fulani, like the nomads, pasture their herds in dry pastures around wells in the dry season, whereas others during this season send their cattle to graze on the *burgu* grasses created by the flooding of the inland river deltas, especially that of the Niger. In what follows I will for the most part treat both major modes of semi-sedentary life together, and will point out regional and ecological variants as we go along. I will not discuss the sedentary Fulani in any detail, since their concerns and way of life fall, for the most part, outside the scope of this chapter. I intend to highlight the religious elites, however, because I believe that religion is potentially the most powerful single force operating in Fulani life today. It is certainly the force that is most likely to mobilize people for action, whether it be for good or ill.

By taking the best figures I could get concerning the states of West Africa, I calculate the total 'semi-sedentary' Fulani population to be about 3 018 143. This includes the 210 000 users of the *burgu* grasses of the river deltas. The number of Fulani following the sedentary life-style amounts to nearly 4 million, mostly in northern Nigeria. To this figure we should add most of the Fulani population of Guinea (*c.* 1.5 million) since few Fulani there still live a pastoral way of life. Though the Toucouleur of Senegal and Mauritania do not define themselves as Fulani, they speak the same language; their life-style is definitely sedentary also. If we add in their population of 512 000, this would bring the total number of sedentary Fulani and Fulani speakers to approximately 6 million persons.

It is clear from these figures that from half to two-thirds of the Fula-speaking peoples (depending on how you count the Fulani of Guinea) are sedentary in their way of life. Of the rest, the vast majority follow variants of the semi-sedentary life-style that I shall now describe.

Two major features distinguish the semi-sedentary from the nomadic Fulani: the practice of agriculture and the practice of transhumance. The necessity for transhumance arises as a consequence of combining agriculture with pastoralism. Much confusion surrounds transhumance and thus it will be discussed first. I will be using the term with a specialized meaning, namely the long distance movement of a herd in the rainy season under the guidance of a small number of male herders. Transhumance, in this definition, necessarily entails the seasonal splitting up of families and herds, with one part going on transhumance, the other part staying at or near 'home'.

During the rainy season, both the nomads and the transhumance herdsmen from the semi-sedentary groups are moving about in the pastures of the Sahel with their cattle and it is during this season also that these groups take their herds to salt licks or salt earth areas, which are places where by licking or actually eating the ground the cattle can get a supply of many necessary minerals that are lacking in their usual diet. Both the nomads and the transhumant herdsmen have an abundance of milk, but whereas the nomads use this to feed their whole families and to exchange for things they need, the herdsmen cannot do this because they do not have women with them. The herdsmen milk the cows for their own consumption, of course, but they do not make soured milk or butter for sale. It is possible that this is an economic loss for the semi-sedentary Fulani. The bulk of the people are cut off from their cattle at the very time when their cattle are most productive. On the other hand, the milk is not really going to waste, for it feeds the calves which would otherwise be deprived of some of that milk when taken by humans. The semi-sedentary people who stay at home are not entirely without milk, for they do keep a few milk cows and their calves with them. This system does not give people an abundance of milk, however, and there is some disagreement as to how valuable it is for the health of the cattle (Grayzel 1976, p. 30; Breman *et al.* 1978, p. 14). There is general agreement, however, that the distance covered by transhumant herdsmen is much greater than that covered by the nomads (Stenning 1959, p. 93; Dupire 1962, p. 25; Grayzel 1976, pp. 15–22; Breman *et al.* 1978, p. 6).

Although the semi-sedentary Fulani are numerous and live under a wide variety of ecological situations, we can make one additional general point about their mode of life: that is, during the rainy season the herders tend to take the cattle to regions that cannot be grazed during the dry season due to a lack of surface water. The reasons why some groups take their herds long distances whereas others, such as the Jelgobe, do not, very likely are as much historical as they are ecological or economic. This is particularly true of the Fulani groups that make use of the *burgu*, either extensively or to pass through it. When the Niger is in flood, it is important to be able to get out of the delta and onto the bank one wants to be on in time, and one has to know the cattle paths (*burti*) and where they go. Not only that, the whole delta region has a set of grazing regulations laid down over 150 years ago by Shaykh Aḥmadu, the founder of the Fulani Empire of Macina. This set of regulations, which still has a certain force today, gave precedence to specific Fulani groups in designated areas, fixed a rate of toll to be levied on cattle belonging to foreign groups, and set up herding schedules such that the arrival of different herds on the scene would be staggered. And all of this was coordinated with a setting up of cattle paths and even overnight stopping places in such a way as to permit millet farming and particularly rice farming to go on at the same time (Gallais 1975, pp. 357–62). The *burgu* is unlike any other pastoral region in Africa; it is certainly not an area that nomads can wander around in as is the case with the steppe and savanna. Not only is access restricted, as we have just seen, but also each Fulani village within the delta maintains a pasture area for its *dumti* – non-transhumant cattle – that is absolutely forbidden to foreign cattle. This pasture is called the *harrima* (Gallais 1975, p. 359). Looked at in the context of the mode of subsistence of the semi-sedentary Fulani, the *burgu* is a very effective variation of the normal, dry season pasturage system. It is both more dependable and usually more nourishing for the cattle than the method used by the majority of the Fulani of pasturing the cows on standing hay in the vicinity of wells. Because of expanding rice cultivation, however, its days may be numbered as a privileged area for pastoralists (see Chs 1 & 7).

The religious elites If asked whether they believe in Allah and his prophet Muḥammad, all Fulani today would say that they do. It is possible that here and there a handful of 'animists' remains, but my survey of the literature on the Fulani suggests that although some Fulani have only recently become Muslims, and although there is a good deal of variation in the significance of Islam, both to specific groups and to individuals within groups, being a Muslim is today part of the ethnic identity of being Fulani.

What is the role of Islam likely to be in Fulani development? Part of the answer to this question can be found in the role and functions of the religious elites. One fact of particular importance for understanding development problems is that religion or tradition often backs up practices that have long-range beneficial consequences for the group but which may seem 'irrational' or 'counter-productive' in the short term.

In Chapters 2 and 3 the dynamic, unifying role of Islam in West Africa is emphasized, but here I want to focus on the role of Islam in Fulani society and the important functions of the religious elites, called marabouts in former Francophone countries and mallams in former Anglophone ones. The powerful, unifying force of Islam for the Fulani can be seen both in 20th century religious movements and in features of everyday life. Let me give an example of this. Cooperation, as we usually understand the term, does not exist in most Fulani societies. When we Westerners use that word we have in mind a group of people working together for a common goal, a goal that will benefit everyone relatively equally. Fulani do work together, but when they do so it is to help specific individuals, not the community as a whole. For the Fulani the community as a living entity in its own right does not exist; what does exist is the bond of friendship or kinship that links one person to another, and all help that people give one another aims at strengthening such bonds. What we call cooperation the Fulani see as a kind of coercion, because in working for the common good they are working to help people whom they do not want to help and whom they might well wish to harm. The only times when Fulani do work in what we might identify as a cooperative way are times when they are working 'for God'. When they help build a mosque or when they collectively hoe the fields of their religious leaders there is a joyful and giving spirit which largely replaces the usual suspicion and indifference concerning group projects.

The marabouts (*moodi'bbe*; singular, *moodibbo*) have a number of functions in contemporary Fulani society. Though a number of them have been important political leaders in the formation of Fulani states, they do not hold political power as a group. The major functions here are curing illness, both physical and mental, acting as therapist, marriage counselor, and adviser on religious questions, instructing children in the basics of prayer and other religious observances, teaching more advanced students the fine points of one's own areas of knowledge (such as theology, philosophy, mathematics, or law), and guiding and officiating in the various rituals of the society, particularly birth, death, and marriage. Thus there are marabouts of all sorts in Fulani society. Although some of them are venal, narrow minded, or dull, many are well educated in their fields, curious and broad minded, and very stimulating conversationalists. But the best marabouts are viewed by people as being both wise and somewhat detached from the everyday struggles of life, and they can therefore be looked to for guidance in difficult times.

A marabout with a reputation for knowledge, kindness, and success will usually have a following and may often be the effective principal of a school whose students support him and his family by monetary contributions and by performing most of the hard work, both agricultural and domestic, of his household. It would be incorrect to say that marabouts as a class are looked up to, though nearly all marabouts inspire some fear because a person never knows whether a marabout might have some secret means of harming him.

The implication of these observations for social development among the Fulani is that consultation with marabouts might be a helpful preliminary to undertaking any project in a given area. One reason for this is that the marabouts are generally the intellectuals of the community and might be more able than others to convey to foreigners (such as technical experts of US Agency for International Development (USAID)) how their fellow citizens perceive and feel about particular issues. Secondly, if a project had the support of the marabouts it would have a good chance of at least getting a fair hearing and careful attention from the people whom it was supposed to benefit. Finally, if the project required cooperative work or action, a religious emphasis might be more successful than an approach that appealed primarily to reason, self-interest, or fairness.

The Fulani and development

Fulani responses to stress and change A key personal quality in the Fulani sense of who they are is a strong sense of honor and shame. Not only do the Fulani believe they have this quality to a greater degree than their neighbors have, but the neighboring peoples themselves willingly concede the point. A corollary of this high sensitivity to honor and shame is an attitude of superiority and haughtiness toward almost anybody who is not Fulani. This is not at all a kind of group pride. Instead, it is pride in oneself and in one's relatives. A Fulani has no feeling for the Fulani as an ethnic group, but he has pride in himself because he can uphold the standard he was given by being born a Fulani. Interestingly enough, this superior attitude toward members of other ethnic groups does not seem to carry over into attitudes toward the natural world and especially toward cattle. The Fulani attribute success not to their own efforts, but to God and luck. For a Fulani, the major goal is to increase one's livestock and to have many children, and both of these depend on what we would call 'natural' forces, namely the fertility of women and cattle, and the ability of the young to survive. Fulani would not deny the value of hard work. They consider failure to cultivate a field or to take proper care of cattle as sheer stupidity. But in their experience there is just no direct relation between the amount of work one puts in and the reward in number of cattle or children.

My point is that actual economic dependence on cattle is greater among the nomadic Fulani than it is among the semi-sedentary ones, but both groups *perceive* their dependence to be very great. This perceived dependency has led to attitudes among the Fulani of profound respect for their cattle. This is expressed in everyday life in a number of ways. One of these is to never show greediness for, or undue interest in, the products of the cow. To do so is tantamount to viewing the cow's products not as a gift but as a commodity that one controls. Now, from a strictly economic point of view, the cow's products are commodities and the Fulani do control them. Can it be, then, that the Fulani do not perceive this basic economic reality? I believe that Fulani could be led to understand this viewpoint, but that for the most part it is not the relevant one for them. In most areas, raising cattle is not a money-making proposition, but rather a nonprofit, people-and-cattle-making one.

Another manifestation of Fulani-perceived dependence on their cattle is their belief in the superior intelligence of their cattle, which contributes to both their wellbeing. In particular, they believe their cattle are more sensitive than other animals to what is best for them. Unlike sheep and goats, for example, cattle can leave camp in the morning on their own, wander around all day, and come back at night. Cows that have previously made the transhumance to the salt earth area have been known to leave on their own if their masters do not take them there in time. Cattle are sensitive to dangerous animals and their fine sense of smell can also detect the location of water when rain has fallen great distances away. Finally, each cow knows its own name and responds individually or together with the rest of the herd to its master's calls. Melle Dupire writes that *Wo'daa'be* herdsmen can use their responsiveness as a weapon, both for attack and especially for defense:

> It is an astonishing sight to see the demonstration. The herdman takes off at a run, calling his zebus; they follow him at a faster and faster pace, then, at his signal, they stop and surround him. Protected by a half-wild herd that will obey him alone, he can defend against an outside attack. (Dupire 1962, p. 97)

How does this whole system respond to stress? In one sense, we have already seen this, for we have emphasized that a fundamental quality of the Sahel environment is its riskiness. Now we can spell this point out just a little more. The key to the adaptation of both nomadic and semi-sedentary

Fulani to the environment is mobility. The specific organization of herd movements varies from group to group and from region to region, but in all cases these movements enable people to get energy – in the form of milk and herd growth – that would otherwise be lost. Perhaps because the basic adaptation is so successful, it is applied to many other situations of stress, including serious famine, family conflicts, population pressure, political struggles, and war. From our Western perspective we might be tempted to think the Fulani are always 'running away' from their problems, rather than facing them. For the kinds of problems the Fulani face, however, leaving a place and striking out on one's own can be a true solution.

Another Fulani response to stress is resignation. Just as they do not believe their work has much effect on whether they succeed in life, they regard failures and disasters as being caused by forces over which they have no control. It is important to understand the attitude of resignation, because it, too, has its positive side. Though resignation, or fatalism, may hold people back from taking active steps to change a difficult or disastrous situation, it also lifts from them the terrible psychological burden of feeling responsible for that situation.

If I am correct in my impression that the semi-sedentary Fulani are more resigned in their sense of what they can do than the nomads are, this difference in attitude supports the point I made earlier about the drastic changes that colonial rule brought about in the semi-sedentary life-style. The nomads, as we saw earlier, benefited from colonialism because they could now move where they wished without fear of their cattle being stolen or confiscated. Colonial rule thus enhanced their all-important capacity to move. But although semi-sedentary Fulani also benefited from this enhanced mobility, they lost the possibility of a third response to stress that had been important to them, namely the ability to recoup losses or increase their wealth by cattle rustling and by warfare. Today, I believe, *there is no decisive action a young man can take if he wants to increase his herd*. Working harder is not the answer, for he is already herding the cattle of many people and there is no guarantee that the work he puts in will benefit his cattle more than those of his friends and relatives. Before the colonial takeover, however, the decisive action of a raid was possible. The reward of success was great, because one's newly increased herd was not only legitimate but admired. And if one died in battle, the anticipated reward was also great, namely that of being remembered and celebrated in song and story. The importance of this cannot be overemphasized. Everybody knows the stories about the heroes of the past. Even for people born long after those days were over, that model for action is the one by which they judge how they are coping with life today. The point here is that the closing off of one of their major channels of action, combined with the loss we mentioned before of their slaves and of their political autonomy (the colonial regime tended to arrest or depose independent-minded chiefs), has undoubtedly increased the Fulani sense of helplessness and the feeling that God alone can help them now.

Where do we go from here? Appropriate strategies for the dilemmas of development

Problems and pitfalls It is apparent from the preceding pages, I think, that the most massive problem the Fulani face, along with the rest of Africa, is simply the continued impingement of the world economy and Western technology on their traditional way of life. The result has been a rapid increase in both the human and cattle populations. In more down-to-earth terms, population pressure means increased competition for land, both among pastoralists, and between pastoralists and farmers. It means that any augmentation of production may not be available for export, but will be absorbed by the greater local need. On the other hand, it means that the demand for beef will continue to rise and that herders will be able to get higher prices when they sell. However these factors play themselves out, the Fulani are reluctantly a part of the modern world and have no choice but to adjust to their situation. The most that USAID can hope for, I think, is to facilitate that adjustment by helping to ease the painful aspects of it, and especially by trying to see to it that human energy does not get wasted nor natural resources destroyed.

In almost all the projects among the Fulani that USAID is likely to be involved in (see Appendix) there lurks a serious dilemma that will sometimes be evident, sometimes hidden. There is a basic conflict between the Fulani and the state. In Upper Volta, for example, cattle bring in 70% of the country's export revenues (Gallais 1972b, p. 368). But, as we have seen, the Fulani are not generally interested in beef production; they are interested in Fulani production. The question is, then, are there ways of increasing beef production that do not lead to diminished milk production or other undesirable consequences?

I would like to point out two areas in which undesirable consequences might well arise. The first relates to the division of labor between men and women in Fulani society. If an increase in beef production meant a decrease in milk production, this would have serious effects on the husband–wife relationship and on the viability of the family as a productive unit. The second relates to commercial production of milk and beef. When we speak of 'beef' or 'milk' we are in both cases looking at the *commercial* end product from the viewpoint of people used to living in a market economy. But the Fulani still live largely in a subsistence economy, so that from their point of view the choice is not between milk and beef but between milk and money, a commodity that is nearly useless to the Fulani in the bush except to buy more cows. It is in the city that one can exchange money for goods, such as fine clothes, yet these are valued not for their usefulness or quality, but because they symbolize leisure and make the owner a center of attention.

This is the model of wealth that the Fulani know. It is exemplifed by chiefs, important marabouts, prosperous traders, goldsmiths and, nowadays, civil servants. It is essentially a sedentary life-style that is radically different in spirit from the way of life lived by the pastoral Fulani in the bush. Though pastoral Fulani would generally say that such a life is not for them, a few cannot resist the temptation to seek its pleasures and buy the goods that symbolize it. They sell their cattle, sometimes other people's as well, and disappear more or less permanently into the cities. Such people almost never achieve sufficient success there to return home without shame. This serves to insulate pastoralists from city values.

A key cultural factor that helps maintain this gap between pastoral and sedentary systems of value is the pastoral notion that cattle are like an investment portfolio. Thus, although it is possible to convert this 'portfolio' into cash, there is an extreme reluctance to do so. It is significant to note that the normal way of saying 'to spend money' in Fula is to say that you 'eat' the money. To say of someone that he 'ate' money is invariably a critical remark not only because it implies that he was self-indulgent, but also because eating itself is somewhat shameful in Fulani culture. It implies a giving in to base human needs that is unworthy of a true noble. Thus there is strong social pressure against selling cows not because they are prestige items, but because to do so diminishes one's capital and at the same time reveals one as 'eating', i.e. giving in to a weakness unworthy of a Fulani.

I dwell on these cultural facts because they have serious implications for economic development among the Fulani. One is that keeping traditional Fulani values alive is crucial for the Fulani to maintain their herds in the difficult Sahel environment. Everything depends on not giving in to the temptation to 'eat' the cows. Encouraging the Fulani to become oriented to beef production, however, inevitably means increasing the temptations to sell animals. This could be highly dangerous unless valid alternatives to 'eating' the money were available.

To put the conflict in a slightly different way: the pastoral nomadic and semi-sedentary way of life is hard; therefore its practitioners must have rewards and *those rewards must be usable while continuing to live as a pastoralist*; money, today at least, does not fulfill this condition. As a consequence, herding is on the decline. My study revealed general agreement among both old and young people that people did not 'love' cows as they used to. This was evidenced by the fact that fewer young men were willing to go on transhumance than in previous times (not just the 'old days', but even 10 years before), and that cattle were allowed to wander every day in the bush untended. It seems that even now the old and the new life-styles are so antithetical that it would be vain to hope they could blend harmoniously.

Building on Fulani strengths We have seen that up to the present the nomadic and semi-sedentary Fulani have been living a mode of life that is particularly well adapted to capitalizing on the riskiness inherent in the Sahel environment. It is a life of considerable physical hardship and discomfort, with the additional stress for those who practice transhumance of living away from the family (and usually without women) for long stretches of time. The Fulani have made virtues of the ability to face these hardships; they have evolved a number of social devices that help them both to resist the temptation to sell off their cattle and live more lavishly, and to avoid any kind of dependence on Western products or the Western economies in general. Thus we are confronted with what appears to be a paradox: one of the major Fulani strengths in a development context is their conservatism. We noted in the Introduction the Fulani resistance to education. Although that resistance may have bad consequences for them, it is also part and parcel of attitudes that are necessary for *good* herd management. Although the Fulani definitely believe in individual ownership of cattle, their fundamental attitude toward them is more like that of a trustee watching over a trust. We have seen that in fact the cows tended by a herd manager do not all belong to him, and even his own cows are viewed as belonging partly to the inheritors who will eventually receive them. That the Fulani tend to place the health and safety of their cattle ahead of their own ease and comfort is a definite asset in the time of severe stress that the economies of the Sahelian countries are now entering. Development programs that will most benefit the Fulani and the countries they inhabit are the ones that build on, rather than attack, these basic Fulani values.

Another strength the Fulani have is the ability to recognize a problem and take concerted action to deal with it in a traditional framework. We saw that cooperation in our usual understanding of the term is foreign to Fulani thinking, but there are other modes of concerted action. The difficulty with these modes is that they tend to have political, religious, or military overtones which are either threatening to governmental authority or illegal in the context of national order. The most striking example of this sort of 'self-help' is the pastoral code of the Dina, set up by Shaykh Aḥmadu and the Great Council in year three (1821) of the Dina, in the Fulani religious empire of Macina. The aim and largely successful accomplishment of this code was to organize the movement of cattle and other ruminants so that the delta region of the Niger could support both pastoral and nonpastoral economic activities such as millet farming, rice farming, and fishing. Areas for pasturing in the *burgu* and for farming were clearly delineated; migration tracks were given fixed locations, as were stopping places and river-crossing points. Shaykh Aḥmadu wanted the Fulani to be more settled so that they could pay more attention to God, so he gave to the transhumance practices a quasi-military organization which enabled the majority of the faithful to remain at home while a small number of young men took all but a few milk cows to distant pastures and salt earth areas (Republic of Mali Ministry of Production 1972, pp. AII/4–AII/6; Gallais 1975, pp. 358–9).

Other examples of traditional action that should be built on, though seemingly insignificant on a broad scale, are driving off diseased cattle from wells so as not to spread the infection to other cows, policing the bush to prevent unauthorized cutting of tree branches to feed goats, and putting out a bush fire and punishing the people who set it. This last case is interesting because it illustrates the bind the Fulani are in with respect to the authority of the state. Traditionally, once the fire was under control, the older men would form a council and, if the evidence warranted it, they would all go and 'eat' the person responsible for it. This means that all the men pay a visit to the offender and, as a good Fulani, he must offer food to his guests or suffer great humiliation. To feed such a large delegation would be equivalent to paying a large fine, since he might have to slaughter most of his animals. In the case I witnessed, some argued the government would crack down on the 'eaters' if they followed this tradition and that they should do nothing until the authorities had been notified. This recommendation finally prevailed. The gendarmes came and arrested the youths who had set the fire (through negligence) and let them off after a light fine.

Finally, the most recent and innovative Fulani initiative that surely must be supported is known as *Laawol Fulfulde*. This institution is being built up by most of the nomadic and semi-sedentary Fulani living near the left bank of the Niger River south of a line from Tillabery to Filingue; its name means 'the way of Fulani wisdom'. French-speaking Fulani translate it as: 'La voie de l'education peul' – The way of Fulani education. The aim of *Laawol Fulfulde* is to formalize traditional patterns of lending out animals, to build an institution that can enforce the obligations these loans entail, and to develop a spirit of cooperation based on religious feelings and on traditional Fulani attitudes towards cattle and herding. It is based on a custom called *nanga na'i* or *ha'bba na'i*, 'grab cows', or 'tie up cows', which is common to both nomads and semi-sedentary Fulani in the Niger–Nigeria border region.

This custom, as described for the *Wo'daa'be* for instance (Dupire 1962, pp. 136–8), involves only the contracting parties and their descendants who continue the relation, whereas *Laawol Fulfulde* is a kind of club which herders join voluntarily and which not only enforces the obligations of men who have such a loan between them, but also acts to ensure that no harm comes to the cattle of members and that members help one another in time of need. For example, if a member sees that a fellow member's cow is hurt, or has strayed into a field, and does nothing about it, then *Laawol Fulfulde* will fine that person for his inaction. Similarly, if a man's wife is sick and he wants the herding group to put off its departure until she is well, the group is supposed to wait. But if one person leaves anyway, *Laawol Fulfulde* will fine him also. These are but a few of many situations dealt with (Centre Régional de Recherche et de Documentation pour la Tradition Orale 1969, 'Ha'bbanaaji', pp. 12–16).

Summary

At this point I want to make one observation and propose one course of action. A major cause of the failure in large development projects is that the people who are supposed to benefit from them rarely participate in planning or executing them. This means that the experts lack information of two kinds: (a) they lack the pastoralists' own detailed knowledge of their milieu and their animals; and (b) they lack insight into how pastoralists would react to or make use of the project's installations. This lack of information is partially due to a reluctance on the part of experts to communicate with the Fulani. The latter, in turn, seriously lack information that would help them make better decisions concerning everything from well-digging to services available to them. One of the best things USAID could do would be to create channels of communication so that more complete and more relevant information could reach the Fulani.

The way I suggest that this be done is through people who would be somewhat like county agents in the United States. They would also be part teacher and part 'group facilitator'. Their job would then include both informing the Fulani of world events and trends that are relevant for their concerns, and learning from the Fulani what they see as their greatest problems. After such a person had lived somewhere long enough to become familiar to the people, for example, I can well imagine that he might get the local *imam* to help him organize meetings at the mosque after the Friday prayers where such information and points of view could be exchanged. People already use these occasions to talk among themselves about current issues. The 'facilitator' might also have a more active function: he could prod people into thinking about and discussing problems that they had not yet fully faced or noticed, and he could propose to them solutions or ideas that had been tried or thought up in other places. In addition, if a road, dam, or abattoir was to be built, such a person would be extremely well placed to learn how it is viewed by the Fulani and thus how the Fulani would react to its installation. At the same time he could help the local population work out a fruitful adjustment to the changes it would create in their lives.

Appendix: selected examples of projects tried among the Fulani

This selection of projects is necessarily limited and personal. I do not pretend that it is exhaustive or up to date. It does include, however, the major areas of life in which development projects have been tried.

The Djibo butter coop In the 1930s a French Commandant de Cercle tried to set up a butter cooperative in Djibo. This was shortlived for several reasons: (a) the people were simply ordered to bring in their milk and most of them never saw any return; (b) there were nearly insurmountable problems of refrigeration and transport; (c) because of seasonal fluctuations in supply and growing refusal of people to participate, it became impossible to assure a constant flow of butter to the market.

The Markoye ranch The original purpose of this AID project seems to have been to demonstrate good range management techniques to the local population and to make available to them breeding bulls for the improvement of their own herds. I do not know whether local herders made use of these bulls or not, but the ranch simply could not be a going concern because it was too expensive. Even with enough water, the ranch did not represent anything the Fulani could reasonably learn from. They did not need to be told that grass would grow if cattle did not eat it, and the fencing in of the range is not only prohibitively expensive but contrary to Fulani custom.

Wells and boreholes Water has been perceived by everybody as the major area where help is needed to improve life in the Sahel. One of the advantages of a well is that once it is set up it remains quasi-permanent, serving the people for years and years. Generally speaking, the desired goal is to provide a more sanitary water supply for the people. But the situation is quite different in the case of wells for the watering of animal herds. We saw earlier that among pastoral peoples each well belongs to the person or group that digs it or commissions it. An important ecological side effect of this social arrangement is that only the cattle belonging to a restricted group of people will normally water at any given well; this limits the number of cattle in a particular area during the dry season and thus helps ensure that there will be enough standing hay to feed those cattle until the rains come. Although pasture is not considered to belong to persons or groups, this arrangement leads to the same result as pasture ownership would. Thus the system of well ownership is a kind of unconscious range management practice.

What is the effect of establishing a public well for watering cattle by hand or a pumping station that brings up the water by diesel power? In both cases, but especially the latter, such a concentration of animals arrives at the wells that the pasture all around them is soon used up and their water thus becomes unusable for the rest of the dry season. Not only that, before this extreme point is reached a law of diminishing returns operates such that the effort saved in not having to extract the water by hand is effectively wiped out by the effort it takes to bring the herd from the pasture to the well and back again, not to mention the effort of getting and transporting water for human use over 10–15 km. This has been observed both in eastern Senegal and in Niger, where the major projects have been tried.

Edmond Bernus analyzed the consequences of setting up a pumping station at In Waggeur in the territory of the Illabakan Tuareg northeast of Tahoua (Bernus 1974a). He reported that, prior to 1948, when the only source of water was a seasonal pond, the Illabakan stayed in the region until the pond dried up (between October and December), but in 1961 when the In Waggeur pumping station was set up, they could stay in the region all the year round. Their region also attracted several other Tuareg and Fulani tribes. Soon the number of animals watering there was triple the number originally planned to maintain the balanced pasturage; at other similar stations the figure was at least double the figure originally planned and in one case (Abalak) it was four times that figure. The Illabakan were unhappy with this development and they eventually requested the administration to turn off the pump. Some families even moved away to the edge of the denuded zone and started to dig their own wells by hand. In 1971, perhaps because of the Illabakan's request, the In

Waggeur pumping station was closed. That year, the Illabakan put their old well back into operation and were able to pass the dry season near it. 'All the tribes that had come these last years had deserted the region, and the Illabakan were the sole masters again, as they had been nine years before' (Bernus 1974a, p. 125).

The immediate problem was that neither the government nor the Illabakan had an effective way of limiting access to that water, which was extremely attractive both because of its location (in fresh pasture) and because it relieved people of the exhausting task of lifting water 90 m. One reason why access could not be limited is clearly that the well is an anomaly in being *a dry-season source of water that belongs to no one*. There were no precedents in custom for dealing with that situation. And truly serious thought and discussion by the Tuareg about this anomaly were precluded by uncertainty as to whether the machine would really stay in operation or what the state was going to do. Most machines in West Africa eventually stop running and are not fixed or replaced; people are not going to adopt a major change in their life-style for anything so precarious.

Introduction of chemical fertilizers, fungicides, and pesticides Though avoiding the usual pitfalls associated with the introduction of an innovation, efforts to promote the use of chemical pesticides by the Fulani have been unsuccessful. Why? For one thing, the extension worker never actively sought to learn how the Fulani perceived what he was doing and the products he was selling. Thus he had little awareness of what their objections might be and did not actively try to deal with those objections. The main objections centered around the fact that these products were powerful poisons. This frightened the Fulani because they thought their animals might accidentally eat some. In addition, the Fulani have little storage space to keep such products away from children. Finally, some Fulani even believed that the poison would be used to poison one another's cows or wells. I am convinced that fear of being regarded as up to no good dissuaded some Fulani from trying these products.

Legislation of pastoral and agricultural zones In most Sahelian countries there has been a gradual encroachment of cultivated fields on grasslands traditionally used by the pastoralists. This has undoubtedly been caused by the pressure of population growth combined with the inability of pastoralists in the colonial and postcolonial periods to prevent the immigration of farmers into their territories. This 'encroachment' is serious because it often amounts to farmers carving out their fields in a dry season pasture area near a well. When this happens, precious pasture reserves for the dry season can be lost (see Barral 1967). The problem is the blocking of herd access to the water. The most common way of conciliating the conflicting modes of land use is to insist on keeping open a corridor in the millet fields allowing access to the water. Often this is handled at the local level rather than by national decree (see Horowitz 1975, p. 398). Niger has established laws with the express intention of limiting the extension of cultivated fields into the pastoral zones, but these have been ineffective in practice because a few good years of rainfall attract the farmers beyond the established frontier (Bernus 1974b, p. 141).

Cattle vaccination programs All the Sahelian countries received considerable technical and economic aid to carry out the Joint Rinderpest Campaign of the 1960s. Fulani and government officials alike agree that the vaccination campaign has been successful and beneficial. On the other hand, there is some evidence that this and earlier veterinary interventions may be detrimental by contributing to a rapid growth in cattle population with which human institutions, particularly in the severe stress of the 1968–74 drought, were unable to cope. A similar problem developed after a 'successful' campaign to eradicate diseases from the cattle of the Karamojong of Uganda. The campaign successfully removed an important check on the growth of the cattle population, yet no steps were taken to provide additional grazing land for the increased herds (Baker 1975, p. 195). Furthermore, the growth rate of the herds after the campaign was reported at 5% per year, which means that the cattle population was doubling in about 15 years (Baker 1975, p. 195). The growth rate of cattle in the *burgu* region of Mali between 1957 and 1970 was also about 5% per year (Republic of Mali Ministry of Production 1972, p. 66), which means that we can expect herd size to increase rapidly and more and worse famines in the future.

Human health programs If the human population were constant, or growing significantly more slowly than the animal population, then it might make sense to speak of reducing herd size. The pastoralist population, however, is growing very rapidly, though somewhat less rapidly than is the sedentary population in West Africa. Gallais reports that the annual rate of population increase among the Bororo is 11%, and the average for all the sedentary populations of Niger is 25% per year (Gallais 1972a, p. 307). Like the growth rate of the cattle population, the rapid increase in the human population is due to a combination of 'peace' established during the colonial period and the widespread, successful health programs to eradicate diseases such as smallpox, measles, tuberculosis, leprosy, and sleeping sickness. No provisions were made to accommodate the increase in population. Birth control programs, for example, have not been tried anywhere and under current conditions there is not a ray of hope for them even if they should be proposed. There are two basically cultural reasons for this that seem consistently overlooked by the World Health Organization and other authorities. One is that every person's success in life, both economic and social, depends on having children. Secondly, to interfere with gestation and birth appears to the Fulani tantamount to trying to control an essentially divine process. The only rural Fulani I ever met who had the slightest interest in birth control were occasional women who wanted some way of stopping their rivals or co-wives from having children.

'The Lord giveth and the Lord taketh away' is what I have shown as the Fulani attitude of resignation, which has important psychological and social functions in society. This attitude will continue until Fulani are accorded greater political independence in their host countries. In the meantime, birth control programs or any other health programs that require the Fulani to take an essentially secular, 'rational' attitude toward life processes and the phenomena of health and illness will at best be reinterpreted according to the Fulani view of how the world works, and at worst be misapplied or totally rejected.

References

Baker, R. 1975. 'Development' and the pastoral people of Karamoja, Northeastern Uganda. An example of the treatment of symptoms. In *Pastoralism in tropical Africa*, T. Monod (ed.). London: Oxford University Press.

Barral, H. 1967. Les populations d'éleveurs et les problèmes pastoraux dans le nord-est de la Haute-Volta (Cercle de Dori – Subdivision de l'Oudalan) 1963–1964. *Cahiers de l'Office de la Recherche Scientifique et Technique outre mer, Sér. Scis Humaines* **4**, 3–30.

Bernus, E. 1974a. Possibilités et limites de la politique d'hydraulique pastorale dans le Sahel Nigerien. *Cahiers de l'Office de la Recherche Scientifique et Technique outre mer, Sér. Scis Humaines* **11**, 119–26.

Bernus, E. 1974b. L'évolution récente des relations entre éleveurs et agriculteurs en

Afrique tropicale. L'exemple du Sahel Nigerien. *Cahiers de l'Office de la Recherche Scientifique et Technique outre mer, Sér Scis Humaines* **11**, 137–43.
Breman, H., A. Diallo, G. Traore and M. M. Djiteye 1978. *The ecology of the annual migrations of cattle in the Sahel*. Unpublished.

Centre Régional de Recherche et de Documentation pour la Tradition Orale 1969. *Fulfulde. La voie de l'education peul. Buubu Ardo Galo*. Niamey, Niger.

Dahl, G. and A. Hjort 1976. *Having herds: pastoral herd growth and household economy*. Stockholm Studs Social Anthrop., no. 2. Stockholm: University of Stockholm.
Dupire, M. 1962. *Peuls nomades. Étude descriptive des Wo'daa'be du Sahel Nigerien*. Travaux Mems Inst. Ethn., no. 64, Institute of Ethnology, Paris.

Gallais, J. 1972a. Essai sur la situation actuelle des relations entre pasteurs et paysans dans le Sahel ouest-africain. In *Études de geographie tropicale offertes à Pierre Gourou*. Paris–La Haye: Mouton.
Gallais, J. 1972b. Les sociétés pastorales ouest-africaines face au développement. *Cahiers Études Afr.* **12**, 353–68.
Gallais, J. 1975. Traditions pastorales et développement: problèmes actuels dans la région de Mopti (Mali). In *Pastoralism in tropical Africa*, T. Monod (ed.). London: Oxford University Press.
Grayzel, J. A. 1976. *Cattle raisers and cattle raising in the Doukoloma Forest Area, October, 1974–September, 1975*. Manuscript for the US Agency for International Development and l'Office Malien du Betail et de la Viande.

Horowitz, M. M. 1972. Ethnic boundary maintenance among pastoralists and farmers in the Western Sudan (Niger). *J. Asian Afr. Studs* **7**, 105–14.
Horowitz, M. M. 1975. Herdsman and Husbandman in Niger: values and strategies. In *Pastoralism in tropical Africa*, T. Monod (ed.). London: Oxford University Press.

Jacobs, A. H. 1975. Maasai pastoralism in historical perspective. In *Pastoralism in tropical Africa*, T. Monod (ed.). London: Oxford University Press.

Republic of Mali Ministry of Production 1972. Rapport de synthèse, Annexe A: Aspects sociologiques des conditions de l'élevage en région de Mopti. In *Projet de développement de l'élevage dans la région de Mopti*. Paris: Société d'Etudes pour le Développement Economique et Social.
Riesman, P. 1977. *Freedom in Fulani social life. An introspective ethnography*. Chicago: University of Chicago Press.

Stenning, D. J. 1959. *Savannah nomads*. London: Oxford University Press.

Contributors

Lansiné Kaba is Professor of History at the University of Minnesota. He received his PhD in history in 1970 from the Northwestern University at Evanston. He has studied at Lycée, Henri IV and the Sorbonne, University of Paris. His research and teaching interests are in Islamic reform, political institution, and development in West Africa. For his book, *The Wahhabiyya: Islamic reform and politics in West Africa, 1945–60*, Kaba received the 1975 Herskovits Award for Outstanding Contribution to African Studies.

Paul Riesman is Associate Professor of Anthropology at Carleton College, Northfield, Minnesota. In 1980, his PhD in ethnology was conferred with distinction by the University of Paris. In addition to his work at the University of Paris, he has studied at Harvard University. Riesman also studied at the École Pratique des Hautes and the École Nationale des Langues Orientales Vivantes, and later spent 2 years in Upper Volta, West Africa, studying the Fulani. In 1977, his book, *Freedom in Fulani social life: an introspective ethnography*, was published by the University of Chicago Press.

Earl P. Scott is Associate Professor of Geography at the University of Minnesota. He received his PhD in geography in 1974 from the University of Michigan, Ann Arbor. He has studied at Louisiana State University and Southern University in Baton Rouge, Louisiana. His research and teaching interests are in rural land use, periodic markets, and development in West and Central Africa. In 1976, his book, *Indigenous systems of exchange and decision-making among smallholders in rural Hausaland*, was published by the University of Michigan Press.

Andrew W. Shepherd is Lecturer in the Development Administration Group, Institute of Local Government Studies, at the University of Birmingham, England. He received his PhD in social and political sciences from the University of Cambridge in 1979. He also received his BA degree from Cambridge. He has traveled widely in West Africa, conducted research in Ghana and the Sudan, and has served as a Visiting Lecturer on rural development on the Khartoum Institute of Public Administration course for Local Government Officers, Khartoum. His research and teaching interests are in urban economic planning and rural development, strategies, planning, and management.

Marilyn Silberfein is Professor of Geography at Temple University, Philadelphia. She received her PhD in geography in 1971 from Syracuse University, Syracuse. She has studied at the University of Connecticut, taught at Northwestern University and served as faculty member and consultant to the US Agency for International Development. Her research and teaching interests are in settlement and planned development in East Africa. Silberfein has conducted extensive fieldwork in Africa and is the former chairperson of the Department of Geography. She is currently on leave from Temple University and working in the Office of Multi-sectoral Programs, Agency for International Development, Washington, DC.

Beryl Turner is presently lecturer in Geography at the University of Zambia, Lusaka. Trained in the highest tradition of British Physical Geography, she conducted several years of fieldwork on *fadama* lands. While conducting her study, Turner was associated with the Department of Geography at Ahmadu Bello University, Zaria, Nigeria. Her research and teaching interests are in geomorphology and resource management in the Tropics.

Michael J. Watts is Assistant Professor of Geography at the University of California, Berkeley. He received his PhD in geography in 1979 from the University of Michigan, Ann Arbor. He has studied at University College, University of London and University of Ibadan, Nigeria. His research and teaching interests are in human ecology, social geography, and development. His recent dissertation, 'A silent revolution: the nature of famine and the changing character of food production in Nigerian Hausaland', is essential to an understanding of the link between drought and famine.

Index

For Product Safety Concerns and Information please contact our EU representative GPSR@taylorandfrancis.com
Taylor & Francis Verlag GmbH, Kaufingerstraße 24, 80331 München, Germany

www.ingramcontent.com/pod-product-compliance
Ingram Content Group UK Ltd.
Pitfield, Milton Keynes, MK11 3LW, UK
UKHW021121180425
457613UK00005B/171

* 9 7 8 1 0 3 2 7 4 4 8 1 0 *